America's
Technology Slip

America's Technology Slip

SIMON RAMO

Director, TRW Inc.

Chairman of the Board of Directors
The TRW-Fujitsu Company

Visiting Professor, Management Sciences
California Institute of Technology

JOHN WILEY AND SONS

New York • Chichester • Brisbane • Toronto

Library of Congress Cataloging in Publication Data:

Ramo, Simon.
  America's technology slip.

  Includes index.
  1. Technology — United States.   I. Title.
T21.R24   338.4'76'0973   80-21525
ISBN 0-471-05976-5

Printed in the United States of America

10 9 8 7 6 5 4 3 2 1

contents

1
Gods' Gifts or Devils' Doings?   1

2
Rewards on the Horizon   15

3
A Slip That's Showing   41

4
Efficiency Isn't Everything   63

5
Limited Free Enterprise and Pervasive Government   71

6
The Missing Decision Department   95

7
The Nuclear Energy Stalemate   117

8
The Synthetic Synthetic Fuel Controversy   143

9
Salvation Through Conservation   162

10
Urban Transportation in Slow Motion   177

11
Food Leadership by America   185

12
The Environmental Government–Private Mismatch   198

13
Extending Human Brainpower   210

14
The International Dimension   234

15
National Security and Technology Transfer   250

16
Education for a Technological Society   268

17
What Will Happen; What Should Instead   281

Index   289

1

gods' gifts or devils' doings?

The human species has not changed significantly in many thousands of years. No basic alterations have occurred in the inherent designs of the body and brain. If a live baby from 100 centuries ago could be deposited with a similarly rooted family of today, that child might grow up to be shorter in physical stature than the average or have a different susceptibility to certain diseases, but otherwise would exhibit little more deviation from the rest of the community than many of its members would show from each other.

The societies of humans, their organizations for ensuring survival and providing satisfactions, have been modified over the period of deducible history, but not because of changes in *Homo sapiens.* Social variations have resulted from people's accumulation of experience and knowledge about the physical universe and themselves, and from the use to which they have put this learning. Indeed, the know-how that human beings have accumulated up to the present moment can be described, with only mild fear of oversimplifying, by the words "science and technology." There has been disappointingly little progress in fundamental social dimensions.

Is the democracy of today's world more advanced than that of the ancient Greeks? Probably—in some ways. Yet, what about this century's ruthless dictatorships, the tribal wars fought in the midst of starvation, the repressive and inhumane acts inflicted by many existing governments on their own people? Brutality, fear, hatred, jealousy, distrust of the "other tribe"—these characteristics are discernible even among so-called educated and refined humans. Antisocial responses are inadequately contained by the thin veneer of accumulated

1

behavioral advance; with provocation that need not be severe, they can quickly break out.

Thus it was only forty years ago that a major European nation, one highly developed technologically but not in all ways socially, committed mass genocidal horrors. More recently we could observe that after a mere half-hour of psychological pressure some automobile drivers in America, in disputes over their places in line while waiting to purchase gasoline, drew pistols and shot at each other. Speaking more broadly, we can hardly focus better on the limited social progress of humankind than with a contrast of two facts. First, we learned what atoms are made of and invented how to release such enormous amounts of energy from manipulating their constituents that it has become possible to wipe out most of the earth's population 30 minutes after a decision to do so. Second, we have not been able to follow this technological breakthrough, even decades later, with a degree of social progress that would preclude the possibility of such a catastrophe.

If 100 years ago, a thoughtful visionary group had gathered to conjure up a list of future human achievements to be prized as valuable and inspiring, they might have included among others a specific pair: the ability to walk on the moon, and permanent peace on earth. To the wiser of those engaged in the discussion, both feats might have appeared overly imaginative and impractical — conceivably realizable, but only a long, long time in the future. Startlingly, we have accomplished one. Not so startlingly, the other still remains as before: conceivably realizable, but only a long, long time in the future.

But if the species has changed physically hardly at all and the allowed behavior of humans, singly and in tribes, has been refined only slightly over a thousand generations, our knowledge of the environment, of the anatomy of our bodies, and of the means to put resources to work for our health, material needs, and comforts has recently exploded. In the past two centuries, our skills and tools have advanced more than in all the millions of preceding years. The first scientific list of the known elements was published by Lavoisier less than 200 years ago. His list included about 20 and, building on his giant step in understanding the basic nature of matter, scientists quickly added recognition of more chemical elements so they now number over a hundred. Throughout all of earlier recorded history, however, matter was thought to be made up of combinations of four elements: air, earth, water, and fire.

The changes that have occurred during recent centuries in the organizational and behavioral aspects of human society are almost entirely caused by the burgeoning of practical know-how. People and materiel can now be moved in hours from any point of the globe to any other. Electronics can keep the nations of the entire earth in instantaneous communication. We feed, move, inform, entertain, clothe, heal, and kill each other by means that did not exist two centuries ago. Divide the peoples of the world into two groups, one in which science and technology flourish, and the other in which the scientific approach is unknown and no wide employment of technological know-how takes place, and the societal differences between the two groups will be profound, more significant than the greatest of their dissimilarities owing to race, religion, or geographic location.

Imagine that we were to draw three curves, with time the abscissa for each, stretching over, say, 100,000 years. The ordinate of the first curve is to depict change in the basic makeup of the human species. We would see the curve as virtually flat, the alterations perhaps discernible to a minute degree only by the most expert biologists and anthropologists. The second curve is to measure social advance of the human race—progress in the relationships between individuals and among large and small groups of human beings. Perhaps this curve would have an observable, slightly positive slope designating improvement. But let us get to the third curve, the one that represents scientific discovery, technological change, and the modifications of the way of life and the physical structure of civilization that come from putting science and technological know-how to work. The ordinate would be minuscule, relatively speaking, for all time until a few centuries ago; then, and particularly in the past several decades, the curve would shoot up steeply, almost vertically, and off the chart.

No wonder we are experiencing the malaise, dislocations, and frustrations of an immense, almost uncontrollable imbalance between rapidly accelerating technological advance and lagging social progress. The more socially immature we are, the more difficult is our problem of social adjustment to the still further advance of technology. Our failure to make a harmonious merging of advancing technology with parallel social progress makes us a "disquiparant" society. In the theory of logic, a system is disquiparant if its definable, separate aspects lack a logical connection. If technological developments are not marching in step with social goals, how can there be logic in our employment of technology?

Today a severe mismatch exists between the high potential of technological advance and the low rate of social–political progress. The reason for this mismatch is not science and technology per se. It is rather that our social organization cannot use these tools to the fullest. Critical and controlling are the interfaces between technology and such non-technological factors as setting goals, examining alternatives, and making balanced decisions. These factors are not now being managed, or they are being handled helter-skelter by people who lack understanding of the process. In choosing where and how to apply science and technology in America it would be helpful, for example, if we possessed clear national goals. When we find it hard to articulate and decide on what kind of society we want, it is understandably difficult to pinpoint the effective use of science and technology to help build it. Our society is not an assembly of related, essential components integrated into a harmonious whole.

Satisfactory decisions in any society cannot be made without an understanding of tradeoffs and options. For instance, we should be in a position to compare the good or benefit that can come from specific technological advance against its bad qualities or its cost. As to our employment of technology, we can be likened to a group of inept carpenters. Equipped with strong and ever sharper tools, they use them clumsily, often getting their fingers in the saws, hitting each other's heads as they bring their hammers back and their own thumbs as the hammers come down. They are not sure what they are trying to build. Confused, yet sensing an unsatisfactory situation from which they would like to extricate themselves, they meanwhile blame the tools.

A short while ago we were confident that the quickest, surest route to the better life was to acquire more scientific knowledge and expand our technology base. Scientific and engineering advances were regarded as limitless sources of higher living standards. These disciplines of the human brain were on a high pedestal. If scientific research and technological development were not worshipped, at least the highest level of such activity was revered and encouraged. To put science and technology at our service creatively and efficiently, a melting-pot form of Yankee ingenuity, an assumed innate ingredient of all Americans, was envisioned as available and ever growing. From time to time we might have to suffer a depression, some incompetents or crooks in influential places, a penalizing war or an annoying number of persistent social problems, but we Americans believed we could count always on one strong and favorable characteristic of our country: our advancing technology would stead-

ily originate new and better approaches to meeting every requirement of our lives and would furnish us with continuing physical enrichment.

With more science and technology, we believed, we could do
anything and ultimately would. And why should we not be so persuaded? In seemingly no time at all we had gone from horse-drawn
carriages to automobiles, then to airplanes, with ever higher speeds
and comforts such as four-speaker hi-fi in the cars and movies on the
airplanes. Radio was invented and soon advanced to black-and-
white TV, then color TV, cable TV, and intercontinental TV by satellites.
To the early vacuum cleaners were added electric dishwashers,
garbage disposals, and washer-dryers. We have found ways to collect, modify, and put to our use all matter of which we know the
universe to be composed and have synthesized superior materials
that do not appear in nature. We have created cities of weather-
controlled structures and automated the mass production of physical
goods. We are used to making low-priced long-distance calls by
direct dialing. We have learned to so control insects and fertilize the
ground as to grow far more food per acre than we can consume
domestically. We have acquired nylon pantyhose, shatter-proof glass,
frozen foods, and microwave ovens. A communication satellite weighing one ton now provides more channels of communication than
200,000 tons of cable laid under the oceans. During one year, a
widebody airliner now moves more people back and forth across the
ocean, at higher speed, than the largest ocean liner a thousand times
heavier could carry. One of today's hand-held computers can make
complex computations that at mid-century would have required
equipment filling a room, with the costs proportionate to the equipment weights.

In not too long a time into the future, we have surmised, every
individual will be able to push buttons on a wristwatch transmitter to
call out a digital code that will establish radio contact with any other
chosen person in the world. Soon our telephones should provide
accompanying sight of those speaking, and our home TV should
have a 3D picture. With advancing technology continuing to produce
more for us with less effort, we should go from a 40- to a 30- or even to
a 20-hour workweek. If we could walk on the moon in the '70s, then it
seemed Americans should soon be able to visit Mars. If we should find
ourselves, a few decades from now, in contact with intelligent life
elsewhere in the universe, this will not be regarded as incredible.

Microbiologists have broken the genetic code and begun pinning
down the subtle distinctions between inanimate and living matter. We

can overhaul the human heart and provide artificial kidneys. Vaccines have wiped out smallpox and polio while penicillin has curbed other diseases. Surely, then, a crash program to cure cancer ought to be successful, we assumed. Soon we should complete our conquest of disease, learn how to control aging, and perhaps even be able to use science and technology to alter the human species.

But if it was only yesterday when science and technology were adored as deities by the throngs—we could do no wrong no matter how avidly we applied these tools—it is only this morning we discovered the gods may be devils. An antitechnology wave has broken over us. A substantial fraction of our citizens now suddenly equate advancing technology with evil. As they perceive it, mass production jammed us into congested cities before we learned how to live together. To them TV means vapid, violence-loaded programs that miseducate our children. Gasoline refineries, needed to supply the automobiles, ruin the environment; those same automobiles, they note with revulsion, kill 80,000 people a year, foul the air, and force us to spend hours each day in traffic snarls. They now think we wasted money going to the moon. The atom bomb may destroy civilization and the nuclear reactor may poison the earth. We make our soil more productive but insecticides may do us in. The pill makes it easier to control the size of our families, but it promotes promiscuous sex among the young and is destroying the institution of the family. Many believe the computer is creating an automated robot society in which humans become slaves as digitalized signals and taped responses take over the society and personal privacy and freedom are lost.

It is not helpful to judge those holding these views to be a minority of extremists or, at best, careless listers of negatives who foolishly disregard the vast positives of science and technology. It is now clear to everyone that scientific research and technological implementations won't solve every problem and fill all needs. More powerful military weapons do not guarantee a peaceful world. No technological advances have come along to give us a lasting, plentiful supply of cheap, pollution-free energy to replace dwindling domestic petroleum and counter OPEC monopolies. We truly are impairing the environment and running out of certain resources. These and other facts about the present technological society are matters of legitimate concern. The nation is now aware that dis-benefits generally accompany all human efforts to produce benefits. Any country that understands this and yet fails to compare intelligently the gains and the accompanying harms deserves to have its policies protested. Such criticism is not unpatriotic and should be welcomed as a necessary prelude to realizing reforms that will adjust democracy and make it work in our technological era.

The desirability of carrying on large-scale research and development to accelerate further technological advance has become a controversial issue in the nation. Trends previously assumed to be clearly for the betterment of society are being challenged by new value judgments. Lower GNP (gross national product) is justified, some aver, if that goes with fewer cancer deaths, cleaner air, less congestion and noise, and a life with less pressure. In trying to find the right values, we are being forced to realize that there is no single truth to lead us to them. How do we arrive at limits on our rights to alter the world with technology? We interfere with nature when we dam up a stream or provide a heart transplant or drive an automobile and release its exhaust into the atmosphere or build pipelines or buildings or factories or sewer systems. When is what we do with technology a boon to mankind and when a detriment? Even the smartest people can't answer this query with an all-embracing guiderule. The difference between a wise man and a fool on this question is that the wise man knows that the values on which the answer can be based are not absolute, constant, and unique; they vary with people, situations, and time.

But if the omnipresence of rapid technological development is in part an evil, is it perhaps a necessary evil? Do we have a practical alternative to the technological society? Is it realistic to ban or even to greatly diminish technological advance? Is such a cutback too penalizing to accept for the American society because our values and social structures are so strongly based on a generous availability of the fruits of employing these disciplines? Experienced politicians assure us that no approach to our social and economic problems is politically viable if it contemplates the majority of citizens' accepting a significantly reduced supply of goods and services. It is equally unrealistic politically to expect those now disadvantaged to abandon their aspirations for the higher living standards the majority enjoy. If these are political truths, then the tools of science and technology must be kept sharp and applied vigorously because such action is indispensable for a feasible approach to national problems.

While some want less technological development and more rules to regulate and minimize it, others are pointing to the available statistics and agonizing over evidence that America has developed a serious technology slip. They argue that we are realizing too little of what scientific and technological advance could yield. These advocates of more scientific research and speedier technological development consider that survival of the human race requires these extensions and applications. They think we need to choose between two options: one, a reasonably attractive and safe, albeit not perfect, environment with adequate but not infinite provisions for the human beings on earth; or the other, social instability, deep human misery, collapse of national economies, and wars based on scarcity of resources. They think the choice is easy. The real issue to them is how to use science and technology more fully, not whether to do so.

Of course, when we speak here of putting technology to work fully we do not mean the unthinking application of it, the misuse of technology on projects the public does not in the end really want and that are more harmful than beneficial. Problems of selection and organization arise here. Furthermore, even if we were to attain perfection in the choice and implementation of technological programs we still would not be guaranteed a healthy economy and a happy society. If we handle badly numerous non-technological decisions, we easily can have inflation, recession, high unemployment, wars, and other ills. Without a strong technological foundation, our minimum needs cannot be satisfied. Yet advances in science and technology are not by themselves sufficient. They are merely necessary.

Until very recently, we Americans took for granted that our country is the world leader in technology. This went hand-in-hand with our thinking we have the highest living standards in the world and are first in almost every scientific feat. It is true that some 20 years ago, when the Soviet Union sneakily abandoned the role of a technologically backward nation we had envisaged for it and launched the first Sputnik, our confidence was shaken. However, by sending men to walk on the moon while the Russians were having difficulty merely landing instruments there, we demonstrated to the world we were still champions.

But today we no longer can assume we are ahead. Contrary indications are all about us in the form of European and Japanese cars on our streets and foreign-made television sets and tape recorders in our homes. We are lagging badly in other fields and being overtaken in some areas where we still have a lead. Evidence is building that these trends are the result of some fundamental patterns that cannot be changed overnight. The United States, a country that previously had outstripped the rest of the world in producing goods and services for its citizens, has suddenly become highly concerned about its ability to go on providing a plentiful flow. More than just a handful of pessimists are asking whether the nation's store of resources and systems for deciding and doing things are up to the job of further increasing our living standards or even preserving the present level in the years ahead.

Have we lost our innovative ability and motivation? We enjoyed remarkable advantages over competitive nations in the century now ending. Maybe our organizations and habits of behavior were suited to the past but do not fit the future. Our presently decreased reaping of technological innovations suggests inadequate sowing some years earlier. The total United States expenditures on research and development are a decreasing fraction of our GNP while in Germany and Japan that ratio is increasing. We are investing less in improving our facilities, again as a fraction of our GNP, here off badly from the other developed nations. No wonder our rate of productivity increase has now dwindled to small oscillations around zero and is below that of all other developed countries.

Some Americans believe the world is changing much too rapidly. Others want more and faster change. Everyone must agree technological advance is a forcing spearhead of change and that we are having great difficulty absorbing it. The rapidity of technological advance—because of our inability to adjust to it, realize its benefits, and minimize its negatives—is presenting us with dilemmas and critical choices. If stopping technological advance is not practical because we need the gains it brings, if curbing it is unacceptable for fear of risking the lowering of living standards and security, if the negatives of technological advance and the growing shortage of resources are real and serious, then we are surely stuck. We need to invent new policy-forming and decision-making techniques.

To illustrate, take the stubborn, ubiquitous problem of inflation. Ask the average American citizen which problem should be rated as worse, inflation or the penalties of rapid technological advance, and inflation would surely be named as the curse we most need to eliminate. But the two phenomena, inflation and advancing technology, are related. Indeed, the strengths of technology can be used to fight inflation. In this instance, inflation is the dragon. Technology, sometimes a positive and at other times a negative force, can serve here as the good knight.

To see how, let us first grant that a sound approach to curbing inflation includes limiting the expansion of the money supply. The government must reduce its spending to make possible politically acceptable conditions for curbing monetary growth and as a political practicality must somehow achieve this reduction without greatly increasing unemployment. But the public insists upon the government's supplying health, education, welfare, and numerous other services, and a strong national defense. This vehemence makes the reducing of government spending a near miracle. Under these circumstances, it is hard to exaggerate the importance or the difficulty of arranging for a higher rate of investment in technological innovation. Whether the problem is lowering the cost of national security or increasing productivity, the technological advance route offers real hope for progress. If the voters will not countenance a much lower supply of goods, we must look for ways to increase that supply at the same costs, that is, while using the same human and physical resources. If resources are dwindling, we must learn how to make more resources available economically. There is only one way to have a generally higher standard of living and to enhance the availability of goods and services to those now well below the average partaker. This is by increasing the quantity and quality of what we produce for each dollar of cost to produce it—a natural role for beneficial technological advance.

Scientific research, if avidly pursued, can discover new resources and teach us how to apply the laws of nature more effectively in using all resources. Advanced technology can be employed more broadly to develop economically and socially advantageous products whose manufacture would create new jobs to fight unemployment. Further R&D effort can lead us to superior methods for increasing supply and lowering costs as a counter to inflation, substitutes for materials in short supply, and ways of acquiring raw materials and manufacturing for our needs with less harm to the environment.

Now if, as some say, too much technological development is inundating us with hazards and detriments, then the present United States trend of slowing technology advance compared with some other nations could be a blessing. If we could rightfully equate technological advance to ruination of the environment and to a bad life generally, not a better one, then we should applaud our failure to develop more rapidly technologically. Let the other nations knock themselves out producing more material things, having the mere appearance of gain while actually lowering their real living standards. We, meanwhile, shall rise above such unsophisticated, misconceived, harmful contests.

But such extreme rationalizations will not satisfy the majority of Americans. True, we have all become familiar with the word "ecology." We all know the goals of life are not met by high production totals alone. We must protect our environment and preserve natural resources. On the other hand, we realize we need a plentiful supply of goods, energy, and services. We also feel intuitively that if we lose the ability to provide well for ourselves we are bound to become more dependent on other nations that exceed our performance record. Then both the quality of our life and our freedom to control it will diminish. While we struggle with working out a better match between the potentials of advancing technology and the needs and wishes of our society, inevitable competition with other nations affects us. Total isolation not being practical, our attempts to resolve our dilemmas are influenced by what the rest of the world does.

Imagine for a moment a planet with only two nations, Country A and Country B, each well endowed with human and natural resources. Also assume this two-nation world is a free one—money, products, resources, technology, and labor are allowed to flow freely between the two countries. Suppose that Country A gradually attains a superior, broader understanding of science and technology, is better organized to employ these tools, has greater productivity, and is more innovative. It discovers new ways to use resources, develops substitutes whenever natural resources threaten to run out, lowers the cost of manufacturing and distribution, invents means to diminish pollution of the environment, and continually designs and brings out new products that are socially and economically superior. It is generally more skillful and mature in matching what science and technology make possible to the needs and desires of the population.

With these assumptions, we know what will happen. The citizens of both nations will prefer the products of Country A because they will be cheaper and yet of better quality and more suited to their needs. The industries of Country A will prosper and employment will be high there. The industries of Country B will be in depression and its unemployment will rise. Temporarily, Country B can maintain its standard of living by selling its country's assets, its land and raw material resources, to the citizens of Country A. Country A will amass more capital, some of which will be used for these purchases. Soon Country A will set up, own, and operate plants in Country B. Some of Country B's workers may go live in Country A, where employment opportunities are better. In time, Country B, like an underdeveloped country, will supply lower wage labor for low technology products, descend to a lower standard of living, and be subservient to Country A.

Let us alter our assumptions somewhat toward political realism. As the trend we described begins to be felt deeply in Country B, its citizens probably will elect a government promising to create protective barriers. These will keep out or tax the products of Country A and restrict foreign investment and takeovers. It will subsidize Country B's industry when it is seen to be failing and charge its citizens high tariffs if they insist on buying the superior, foreign-made products of Country A. Country B can isolate itself as though Country A did not exist. The end result, however, will not be much different. Country A, with its advancing technology, will have a rising living standard. Country B, busily engaged in subsidizing its own backward technology industry, will produce less (and lower quality) products for its citizens to divide up.

Some in Country B may argue, "When all is added up we have not lost. We have benefited by not worshipping technology as has Country A. Yes, we produce less, but we have a simpler and better way of life, one that is less dependent on advancing technology." But if Country A has been properly described as superior technologically, it will use technology in a thoroughly optimum manner and the criterion for what is optimum will meet the value judgments of its citizenry. If Country A moves ahead unthinkingly instead and, in producing increasing volumes of products, spoils its environment and impairs the health of its citizens, then it would have to be reckoned as inferior, not superior, in its use of technology.

Similarly, Country B, defined as inferior technologically, is not automatically superior in another sense: it has carefully avoided employing that technology which provides more detriments than benefits. It is one thing deliberately to produce fewer shoes and thereby gain time to walk barefoot on the sands of a clean beach. It is another to walk barefoot because we can afford no shoes— especially if the beach is filthy. A sound definition of technological superiority is not merely to use advancing science and technology aggressively and avidly. It is to select appropriate areas for technological efforts. It is to create approaches that will generate the least negatives and the maximum positives. The objective is not to accumulate the biggest bag of technological tricks, winning a science olympics of discoveries and breakthroughs over other nations. However, if other nations excel on a broad enough front in science and technology, they will be the ones with the most options to set a society pattern of their choice.

How good and how bad for America is further advance of the technological age? By accelerating scientific research and technological developments in the United States, what do we gain and what do we lose and how do the two compare? Is it inevitable that we become an even more technological society? Can we arrange to reap the positives, or most of them, and eliminate the negatives, or most of those, of further implementation of advanced technology? Must we in the United States strive for a position of technological superiority or else lose out to other nations that move faster technologically?

These questions suggest a summary question. Are we in the United States using science and technology to the fullest on behalf of our society? This is not to ask whether we are following up every clue to nature's undiscovered secrets and are building every machine it is technically possible to build. These latter are very different and less sensible questions. We seek here rather to inquire whether our scientific and technological know-how is being applied adequately where there is evidence of high economic and social reward for the effort. If the American society is not now making proper use of science and technology tools, then why not? Are we becoming slower, more timid, and less innovative in applying science and technology? What stands in the way? Is Yankee ingenuity really disappearing? Is there something about the pattern, the rules, the organization of American society that is at fault here? Is our system of applying value judgments, making decisions and implementing them inadequate for the technological society ahead? Should it, must it, be changed?

In the chapters that follow, we shall take up these questions. First, so that we do not base our thinking too much on the past, we shall look quickly at new scientific and technological developments in the offing. We shall observe extremely advantageous potential benefits we shall not want to miss, but also accompanying penalties we should not like to accept. It will be seen that both the government and the private sectors need to be heavily involved in what happens on the science and technology front but it also will be learned that the United States has not yet developed sound ways to assign the proper roles to government and the private sector. It will become apparent that we have not done as well as we must to provide for an understanding of the negatives as well as the positives of technological developments, and that our procedures for comparing the good versus the bad, balancing the two to meet national goals, and making decisions that will lead to the desired implementations demand major alterations. We shall propose better approaches and illustrate them by application to some of our leading technology-related problem areas such as energy, transportation, environmental protection, food and nutrition, and communications.

At this point, having accumulated a long list of tasks ahead for the American citizen, we shall be forced to recognize that success in improving organization and decision making, and indeed in choosing the most sensible societal goals, depends upon understanding by the citizens, so we shall ponder education for the technological society.

We shall then conclude with a chapter contrasting what will likely take place in the period ahead with what, if we persevere to make things turn out well, could instead eventuate. The second scenario, we shall see, is a lot better than the first.

In the competitive and highly interactive world society, advancing technology cannot be halted, but the movement can be influenced. If we do not understand and work at properly employing technological advance, our goals as a nation will not be met. We shall also then not make our proper, needed contribution to world social, economic, and political stability. This will be damaging for America. It will be equally bad for the world.

2

rewards on the horizon

Knowledge begets knowledge. We cannot resist, still less prevent, our continually learning more about nature. Learning engenders doing. We cannot organize society so as to ban putting knowledge to work. And we would not want to. Much that we can avail ourselves of in the future through technological advance is already apparent and manifestly of such great worth we would not elect to pass it up. Unfortunately, in attaining most benefits the distinct possibility exists of reaping some accompanying penalties. If we envisaged these negatives as overwhelming, we would seek to halt technological expansion. Fortunately, such extreme pessimism and action are unwarranted except in individual situations. It is more intelligent to expect that serious side effects will emerge alongside technological developments and then resolve to be so alert, wise, and innovative as to control, limit, counter, and, on a selective basis where necessary, abort specific technological applications.

Advancing science and technology in the years ahead will be replete with breakthroughs we cannot even imagine, so they will surely surprise us when they come. However, apart from developments on horizons too clouded and distant to see, and those amazing discoveries impossible to anticipate, what do future scientific research and technological advance appear to offer us? Granted possible corollary detriments, what can we say today about the way they will stack up against welcomed gains? Let us take examples in frontier areas of science and technology and attempt some reasonable extrapolations. Before we tackle the tough decision-making and organizational dilemmas that society — technology confrontations evoke, let us use these illustrative cases to inquire whether it is all worth the trouble, whether science and technology promise advantages so great they impel us to hone and perfect our handling of these tools.

Some possibilities we shall describe appear to be certain and soon, while others seem more speculative of occurrence and will happen later if at all. Some suggest the realizability of high rewards but only if we invest heavily. A fraction seem to have severe attendant negatives, or will present great difficulties in organizing to implement them. Most require that the respective roles and missions of the private sector and government be worked out carefully because success will depend on cooperation and sensible assignment of responsibilities. All the important future technological advances these examples will illuminate have significant economic–social–political repercussions, good and bad. Finally, that we are not an isolated nation and must relate to the rest of the world will be seen through these examples to complicate matters. This condition, we shall observe, makes it more challenging to garner the gains and curb the potential detriments of the entry of new technology.

We start with an area that may soon move to the top of the list in importance in science–society interactions: food and nutrition. Already today starvation and malnutrition plague much of the world. As the years change to decades, even assuming optimistically a much improved control of population growth, the problem of food shortages may be expected to surpass dwindling energy supply in seriousness.

In the early 1920s, some prominent agriculturists announced that in another 50 years the United States would have to engage in substantial food importing to avoid famine as our population approached the 200-million level. It has turned out instead that our nation today is nearly alone in the world in food self-sufficiency. Virtually every other nation, developed or underdeveloped, produces too few nutrients to feed itself. In 1950 some 40 to 50 nations were either self-sufficient or able to export food. By the middle 1970s, only 20 were in this category and 90 percent of all agricultural exports were by four countries. The United States has a unique combination of land, water, weather, topography, infrastructure, and developed technology in mechanical equipment, fertilizers, herbicides, pesticides, preservatives, storage, packaging, and distribution. As the food requirements of the earth's population increase, our remarkable potential to provide a constantly growing food surplus will be critical to the health and stability of the entire world. Looking at it selfishly, the increasing dependence of other nations on the U.S. supply could place us in a position of foreign trade advantage and positive balance of payments comparable with that of the OPEC nations in oil.

Yet our science and technology today are only at the beginning of a train of advances agriculture and nutrition experts see ahead. From seed to mouth, research and development opportunities abound. Decades ago it took several months for chickens to grow big enough to eat. Then, in the 1930s, chickens began to be raised in confined areas with cod liver oil in their feed and blue light glowing about them to keep them calm. Now it takes only nine weeks to produce a four-pound chicken. It used to take four pounds of feed per pound of chicken. The figure is now down to two and the limit does not seem to have been reached. Through cross-breeding and artificial insemination, chickens and turkeys have been redesigned for meatiness and have attained new highs of efficiency in converting grain to edible protein. Such techniques and other new ones are now applicable to cattle production. Prized cows are fertilized by artificial insemination; then the embryos are transferred to ordinary cows which, as surrogate mothers, carry the prized calves to birth. Researchers believe the full development could yield a four- or fivefold increase in production per cow.

As to basic nutrition science, we have hardly begun to understand what nutrition a human being really needs. The same applies to food animals. Biologists' new discoveries in genetics seem on the threshold of progress that could enable major altering, in fact, actual engineering, of crops to our specifications. These recent breakthroughs also promise future development of superior animals. Farm machinery can be advanced so as to be more effective with less energy consumption. Packaging and distribution techniques can be made more efficient by the application of new computer technology which, applied to the food process, is in its infancy. New ways to protect foods in storage appear promising but need more research. Improved transportation of food through superior technology can diminish storage time and waste. Satellite observations of the earth can yield data which can go directly toward improved control of crop disease and superior agricultural planning for optimum use of land and water. The weather can be predicted more accurately with more research and, in a substantial number of important circumstances, weather control may become practical.

Ill effects of some pesticides, herbicides, fertilizers, and preservatives have been identified and many other health hazards may be expected to surface from these chemicals. We don't want increased food production to bring decreased health. Also, shortages of energy and water may create bottlenecks to production levels. We need to learn how to eliminate all these detriments and limits, but useful approaches are unlikely to be invented without a substantial broadening of the science and technology foundation on which the required developments can rise. The answer is not to ban all chemical fertilizers or herbicides and the rest, or reduce production of food when it clearly is needed by the world. It is rather to advance our know-how. One example would be through research in biological nitrogen fixation and photosynthesis to enhance plant efficiency and crop yield. Next to carbon dioxide, nitrogen, taken into a plant through microorganisms attached to its roots, is the most essential material for plant growth. Most of the nitrogen in commerical agriculture is now supplied by fertilizer but the efficiency is only fair and more than half the applied nitrogen is not usefully absorbed. With more research a genetic redesign of plants is conceivable to increase this efficiency and cut down on fertilizer requirements. Of course, what the scientists would like and expect eventually is to design a commercial plant that takes in nitrogen directly from the air where it is plentiful and do away with fertilizer production.

We are now using only one percent of the world's million species of plants. Experimentation could turn up many advantageous substitutes. Plants now capture only about one percent of the energy in the sunlight falling on them. Increased leaf area would help, but more fundamental advances in the plant's internal mechanisms are necessary (and not scientifically to be excluded) so as to greatly improve the photosynthesis process whereby the radiant energy is used to manufacture carbohydrates and sugars from carbon dioxide and water. It is also to be anticipated that in time plant scientists can design plants with greater resistance to pests, frosts, and other weather extremes, and to salty soil and brackish water.

The payoff from food and nutrition science and technology advances clearly can be impressive for society as a whole, and particularly for America if it is the leader. The possible consequences of not pursuing R&D in this field with determination and effectiveness could include, at their worst, general world political upheavals and instabilities, preoccupation of world leadership with the problems of starvation, larger budgets for armaments, and wars over food supply resources — all well beyond the disadvantage of a mere lower average standard of living.

We note here, although it is getting ahead of our story, that to make the fullest use of science and technology to enhance the American contribution to the world's food supply involves many semiautonomous elements of our population. Scientists and engineers, the universities, small farmers, and big corporations are engaged in the numerous aspects of food production and distribution. Many government units are concerned with research sponsorship, priorities, allocations, subsidies, and regulations on land usage, water supply, food purity, energy, environment, prices, export policy, and more. An enormously challenging leadership problem exists. The government is closest to doing the overall systems integration today, but it is far away from actually filling that role. The government now does regulating and provides research funding but much more needs to be done if, in food and nutrition, the United States is to come somewhere near grasping fully the opportunity to serve society.

Let us turn from ground to sky — to space technology. A satellite in orbit can now provide important economic and social benefits for society on earth. Virtually any two points on the globe can be in instantaneous, reliable communication since the satellite can see around the curvature of the earth. Equipment on the satellites can do more than act as a relay for television, telephony, and other information transmission. Satellites furnish heretofore unattainable observation points from which the earth can be examined in detail and the atmosphere and space surrounding the earth monitored for measurement of numerous phenomena whose understanding is important to accomplishing something useful down below.

The now practical orbital capability, taken together with recent advances in electronics, makes communication over large distances both cheaper and more reliable. Telephony and television between continents, and over large geographical spans of one continent, will save funds and critical materials as compared with other alternatives. In the near future, direct transmission from satellites to the roofs of homes, schools, industrial plants, and office buildings will increase greatly the available channels.

By employing proper equipment in satellites, low-priced and lightweight gear in every airplane in the skies, and computer equipment on the ground, airline navigation and traffic control can be much improved. The signals from a satellite instrumented to be a precision artificial star can be processed so that ground stations will know the location of all aircraft precisely, to figures less than the dimensions of the planes. No further scientific discoveries are needed for this. The system's equipment and procedures only have to receive adequate engineering effort so production and installation of the gear can commence. When satellite-based airplane navigation and traffic controls are installed on a large scale, traffic densities can be higher than now, yet with added safety and lowered cost for reliable control of air travel.

To accomplish this requires substantial initial investment, but the problem is not availability of capital because the return on it would be high. Obviously, neither is the bottleneck a concern about environmental pollution or energy dissipation. In fact, fewer traffic stackups and less expenditure of fuel and air pollution would result along with the first goal, a decrease of accidents. The limiting factor to attain the new system is organization, a task made difficult because of the many groups involved, including governments from cities and regions to nations. The FAA, FAB, FCC, Army, Navy, Air Force, U.S. airline companies, pilots' associations, airport operators, air traffic controllers' unions, satellite builders and launchers, navigation equipment manufacturers, and agents of the airlines and governments of foreign nations operating in the U.S. or where U.S. airplanes land—all are part of the approval or veto loop directly or indirectly. To try to figure out which part of the U.S. government is in the leadership position and has prime responsibility is certain of failure. The respective roles of the various world governments and the 'private sector are also vague. Despite the growing hazards to life at many busy airports, the technology is not being made available to provide service superior from the standpoint of risk to life, economics, reliability of operation, and traffic handling capacity. This potential is not being realized because the organizational task and decision making are apparently too difficult for us to accomplish.

Many other cost-effective satellite systems can be created in the coming years: improved navigation and communication for ships at sea; computer-to-computer communication of the data essential to efficient operation of business and industry; an extensive broadening of education, with a substantial reduction in its cost, by direct broadcasts of thousands of taped video programs to schools, homes, and industry; earth-scanning to improve discovery, analysis, and utilization of mineral, water, energy, agricultural, fishing, and other earth resources; weather prediction, and ultimately the beginnings of weather control; library networks to provide research and operating data to medical, legal, and other professional activities.

Research spacecraft generally offer long-range potential that cannot easily be translated to applications in the near term. However, some data-gathering spacecraft measure physical phenomena bearing more directly on short term economic benefits. For example, satellites can monitor solar radiation effects on the earth and its atmosphere. These measurements then make possible calculations of the energy exchanges between the sun and the earth, which in turn enhance the accuracy of forecasting climatic trends and predicting radiation fluctuations which can enlarge the cold zones of the earth, shorten growing seasons, and shrink arable lands. Use of these data finally can improve planning in land use, transportation, and agriculture.

Signals from satellites six hundred miles up in the sky, processed by advanced electronic systems, are already beginning to help fishermen find fish. Volumes of information about how the earth looks are sent down by satellites each day. An electronic system interprets the data and turns them into super-clear color photos packed with clues about the earth and its waters. These space pictures help reveal ocean features such as sediment layers, depth, salinity, and chlorophyll content. Where there is chlorophyll there is very likely to be the basic food for fish. These photos are extremely valuable to commercial fishing because they can save the fleet from wasting most of its time and fuel looking in the wrong places for its catch.

These earth resources satellites are in their infancy but already are causing new international relations problems. The Philippines, for example, have protested the use of foreign satellites to survey the natural resources and track agricultural production in their country. A coordinated international exchange of information is needed and this is not easy to arrange in view of the diversely perceived interests of the many nations. Another example of the non-technological arrangement-making problems in beneficial utilization of space is the lobbying that has started, to prevent the United Nations' so-called "Moon Treaty" from being approved. The treaty declares that "the moon and its natural resources are the common heritage of mankind." American industry groups are already exploring the profit potential in investment to create gravity-free factories in space and extract and bring back resources from the moon. They ask whether the moon is an eligible arena for private risk taking, and this bears on whether the United States should sign the treaty.

The fact that full use of space technology could produce high economic return on investment through additional, better, or more economical services is suggested by making a rough estimate of the dollar revenues involved. With full implementation of the socially and economically beneficial applications, it is not an exaggeration to expect that by another decade or two around 200 million men, women, and children in the world will be found using a satellite system in some activity for at least one hour a day. Let us suppose conservatively that it is worth around $5 per hour for an individual to be entertained by TV, make a telephone call, obtain data important to his or her work, receive education, or be guided for safe travel in an airplane—all these systems being possible with the right apparatus in space. This adds up to world revenues of a billion dollars a day or hundreds of billions of dollars a year. This is an impressive contribution to the world's gross product, but more importantly, it constitutes a positive lever on the effectiveness and economic value of many of the populations' other endeavors.

Of course, for this magnitude of space technology-based operations to evolve we have to imagine the settling of numerous interface problems between government and the private sector. The radio spectrum has to be allotted carefully and this is certainly a government function. Launching of satellites must take place at government-owned and operated facilities, using booster rockets that were developed with government funds and are available only with the government a participant in the contracting. The government, with its general duties in control of all communications, must design ways to assign privileges to the private sector in space technology and decide what to keep as a government responsibility. Such organizational effort typically moves very slowly in America today because the contests among many interests are hard to resolve. Moreover, essentially every program has requirements for international arrangements, which are even more difficult to negotiate than domestic ones.

From space, let us return to the firm earth and look at an area of technology that might appear at first too mundane to serve as an illustration of future change stimulated by technological advance: the automobile. The relationship of the personal car to the person is now going through an unusual transformation that provides an excellent example of how technological advance can provide benefits if careful control is exerted in implementation. We probably would have to recall the days of the invention of the internal combustion engine, or at least the period when mass production of automobiles was being developed, to match in historic significance the present dynamics of the involvement of people with cars. A number of new issues have come upon us quickly and simultaneously, requiring the automobile to be different from what it has been in the U.S. for decades. The most conspicuous factor now is the imperative to cut gasoline consumption. Control of exhaust pollutants and fulfillment of safety standards are two other design criteria with severe impact on the cars' design and they are here to stay. Simultaneously, newly available technology related to materials and electronics can have a strongly advantageous impact on the meeting of the new requirements.

It might be thought that because western Europe and Japan have long had to contend with high-priced gasoline their approaches to auto design essentially define the needed new vehicles, and that this conjecture is confirmed by the present large scale of automotive imports into the United States. To be sure, a smaller car uses less gasoline to traverse the same distance. But this is only a part of what is needed to bring the automobile of the future and the typical U.S. urban driver into a new marriage more compatible with changing overall national interests.

Materials research is pointing the way to large weight reduction, which goes directly to gas saving. Ceramic materials are being researched for engine parts that are more heat-resistant and lighter than metals. Graphite-based body and power-drive structures, both stronger and lighter, are in early R&D stages. Scientific research on the fundamentals of the combustion process appears to promise higher fuel efficiency and less pollution. The electronics revolution is beginning to provide automotive engineers with means for sensing and beneficially influencing virtually every aspect of the operation of the vehicle from control of flow of fuel and its burning to the governing of power flow and the triggering of safety actions. Electronic microprocessor devices (tiny, versatile, low-priced computers) can now be applied to cut costs of maintenance, save time for the owner of a car, be more stingy in fuel dissipation, increase safety, and decrease pollution. Microprocessors, tied to ingeniously designed sensing devices, can be used to monitor the operation of a car or truck, indicate how many miles to go before gas is needed and when to change oil, rotate tires, winterize the car, and tune the engine, or when the emissions controls need adjustment, or whether the car is being used at optimum efficiency or is being misused.

In particular, engine efficiency can be increased through electronic control by causing engines to respond optimally to variations in the temperature and pressure of air and fuel and in the composition of the fuel. In addition, the auto's sensors and microprocessor devices can be connected with a computer at service stations to provide additional readouts for quick diagnosis and directions as to what to do to make quick adjustments and maintain top performance. Electronics in automobiles will advance considerably beyond today's beginning applications. The brakes can be automatically controlled, when applied, to avoid skidding. The driver can be warned of malfunctions that could cause accidents or unnecessary deviation from optimum fuel economy in fan belts, headlights, engine coolant, battery fluid, door locks, fuse failures, tail or brake lamp failures, low clutch fluid, high exhaust temperatures, and deviation from proper pressure in tires. We can even have electronic aids to prevent a drunk driver from being able to start a car or thieves from getting in and moving it without useful warning signals being generated.

Advancing technology in the automotive field, if pursued vigorously, appears now to offer the potential of providing for a changeover to engine designs using principles and fuels radically different from today's internal combustion engine. We shall cite only one example, the fuel cell. An electric battery soaks up electrical energy imparted to it and stores it for ready availability. A fuel cell, in contrast, changes matter injected into it to electricity. This matter is fuel which is "burned" in an electrochemical sense, producing electrical energy rather than heat of combustion. Fuel cells already have been developed for space vehicles. In the space program, the fuel cells used hydrogen as fuel, combining it with oxygen to make water, a chemical reaction that produces electricity. In a car, the starting fuel might be some form of alcohol broken down by a catalyst to provide hydrogen. The generated electricity would go to a conventional electric motor to drive the vehicle. It is thought by some engineers that a fuel-cell electric car the size of a typical four-cylinder auto of today could eventually be developed to travel about 500 miles at a speed of 55 miles per hour on 10 gallons of methyl alcohol with remarkably little pollution.

Of course, without a lot more R&D, this is just speculation. The fuel-cell idea as applied to the automotive field has so far received perhaps one millionth of the attention the internal combustion engine has had over the last century. With further research it could turn out to make a significant difference in the nation's need for imported oil and in protection of air quality. It is an example of a long list of un-evaluated possibilities. It also illustrates something else. Radical changes in automobile design require many years of effort and tremendous financial resources. Even the first step of R&D on the fuel cell would consume years because so many details must be invented and proven reliable under a broad range of weather and perfor-mance conditions. Emissions that pollute must be understood and properly constrained. The hardware then must be designed for rea-sonable pricing, tried out in large-scale production, and shown to have adequate life. The tooling of factories would use up additional money and time. At the point when all service outlets are ready and mass deliveries can be made, billions of dollars would have been invested. Factors like long time periods, high startup costs, and huge capital investment narrow the main participants in the action to only a few manufacturers and their cash flows are dependent on the timing and severity of government mandated requirements on safety, fuel economy, and air pollution.

Much less speculative and with far greater potential than has so far been tapped is the diesel engine. However, not only is it true that benefits remain to be realized in further use of diesels; some serious potential negatives are surfacing which also need to be explored carefully. Diesel fuel in automobiles and trucks saves oil and money because more fuel can be made from the same amount of crude oil and a diesel yields 25 to 30 percent more MPG (miles per gallon) than gasoline-powered cars. (In cities, where the vehicle is idling or decelerating much of the time, this figure can rise to 50 percent.) But some who have studied the problem fear that diesel powering of cars and trucks may increase the nation's rate of cancer and respiratory diseases. Also, the visibility of urban air could be reduced. Diesel engines produce a complex cloud of exhaust containing as many as a thousand different chemical compounds and some of these are suspect as carcinogenic.

Unfortunately, the chemistry of diesel exhaust and its impact on human health is a rather new subject for researchers. Only in the past few years have scientists even been attacking the problem. The National Academy of Sciences is now preparing for a comprehensive study of the health and safety implications of converting a substantial fraction of the nation's auto production from gasoline to diesel fuel. If not halted, such extensive conversion might take place in another decade. Diesels are cleaner than gasoline engines in carbon monoxide and photochemical oxidants, but they generate far more nitrogen oxides and up to 50 times more particulates (soot) that penetrate deeply into the lungs. Diesel particulates are so complex chemically that scientists have identified less than 10 percent of the total number of compounds contained therein. The chemical makeup varies from batch to batch depending upon the crude oil and the method of refining. This makes much more difficult the problem of determining whether the exhaust is dangerous to human health. The diesel is an excellent example of technological innovation that presents us with both known benefits and unknown detriments.

It may be anticipated that all automobile manufacturers throughout the world will be innovating and applying science and technological know-how with alacrity to create the automobiles of the future. Those that perform best in this technological race will receive rewards of superior return on investment. Their nations will benefit from higher employment in the communities where the plants operate, improved balance of payments, and general economic prosperity and flexibility. In every developed country, the automobile is a key segment of the economy. To be second or third is to accept severe penalties. In the United States, the automobile industry is the nation's largest employer, second only to the government, and yet foreign-made cars now comprise one quarter of the American market. We have 100 to 200 million cars today that preferably ought to be replaced by cleaner and safer models with twice the fuel economy. The total replacement expenditure at the retail level would be around a trillion dollars. Thus, automobile technology is a major arena in the world contest for technological superiority and economic strength. Because of this enormously important fact—not to mention matters of air pollution, fuel economy, and safety, all highly political issues—the government can be expected to be increasingly involved with the automobile industry. Unfortunately, for the same reasons, the government-to-private-sector relationship will remain in contention and this dispute will limit technological advance.

Turning from land to ocean, we note that advancing science and technology make it sensible now to consider the oceans as a source of resources well beyond any expectations we might have had in the past. The oceans have always been used for transport and much of our food is taken from there. Now we are learning to extract oil from ocean-covered coastal land. But there is much more potential. Scientific instrumentation is reaching new levels of versatility, sensitivity, and accuracy, so that we can now observe ocean phenomena, both biological and physical, to provide heretofore unavailable basic data. Also, computers have now attained a range of capability and economy so that the data can be pondered in great detail quickly and the laws of behavior of the oceans understood. This new technology augurs well for our ability to enlarge the utilization of ocean resources.

The establishment of a 200-mile economic zone off our coasts will open up an area equivalent to two-thirds of our land mass. The need today appears to be for systematic exploratory effort at locations and water depths that previously were unavailable for examination but that, with new techniques on the forefront of technological advance, now appear reachable. There are promising regions at water depths greater than 1000 meters. For instance, extraction of mineral resources and hydrocarbon deposits may be feasible below the deep waters on the margins of the continental shelf. Also, both animal and plant life in the seas can be enlarged to serve our needs better. We can expect to increase greatly the protein extractable from the oceans, and ocean farming may become practical.

As to minerals from under the oceans, we should recognize that a serious resource crisis is building in base minerals such as cobalt, manganese, tin, and chromium. For these metals the United States has almost total dependence on overseas sources. The high demand, the frequent foreign efforts to create a monopoly-controlled market, and the chronic political instability of some nations of origin have led to price manipulation, black markets, and panic calls for development of alternatives. For instance, it has been predicted that in 25 years South Africa and Gabon will be the only non-communist sources of manganese. Steel cannot be made without manganese and each of the other metals mentioned is also vital to some industrial products and processes. It may become of critical interest that we learn how to bring these metals up economically from the deep ocean floor where they occur in abundance in nodule form.

The mineral-rich nodules in significant concentrations on portions of the deep seabed involve considerable variability as to composition and concentration, so differing techniques are required to harvest and process them for conversion to useful mineral compounds. Access to the nodules in one region may require investment in equipment substantially different from that needed in another. This makes control of specific regions particularly important to individual investors. In view of these facts, what is our access to the ocean floors of the world?

The United Nations conferences on the law of the sea involve 158 nations. Most of them are quite determined that the deep sea out beyond 200 nautical miles from shore shall be a "common heritage of mankind," this wording coming from a 1970 United Nations resolution. In other words, what is available at the bottom of the sea may not belong solely to the group that gets there first, has the best technology, and extracts it. Ambassador Elliott L. Richardson, who heads the United States delegation to the conferences, has called the treaty toward which the conferences are working far more important than SALT II. The minerals resting three miles down, waiting to be scooped up, have potential value in the trillions of dollars, but it will cost many billions of dollars to capture this treasure. The return on investment nevertheless could be very high. Developing countries are demanding equal votes with the high-technology countries on a proposed international seabed authority. Also, they are insisting on a re-do of the treaty in 20 years with new mining ventures barred until the provisions are renegotiated. And they want an agreement that any companies engaging in activities on the ocean bed will allot half of each mining area to the seabed authority.

All this adds up to strong inhibition of the investment of United States private capital to mine the oceans. Such investment will not take place unless there are clear American rights of access to minerals in international waters and satisfactory rules with regard to exploration, mining claims, and environmental control.

The ocean is also a source of energy. One possibility is exploitation of the temperature difference between the surface of the ocean and the ocean depths. In some ocean areas, for example, near Hawaii, this temperature difference is at a maximum. In time, perhaps, huge power stations may be built as artificial islands generating electricity which then could be cabled to adjacent land. In the process of energy generation, the immediate local ocean temperature would be changed. This and other ecological effects might alter sea life in the vicinity of the facility, perhaps beneficially, perhaps harmfully. Only more research will tell us. Moreover, without considerably more R&D effort than has so far been expended on this approach, the economic feasibility will not be established.

Ocean-thermal energy is just one kind of solar energy, which has become a political and emotional issue. Many among both its most zealous advocates and its insistent detractors have little expertise in the topic, and even qualified professionals are found often in substantial disagreement as to the real potential for solar energy over the coming years. This is because there are differences in hunches as to the likely results of highly accelerated spending with the hope of making discoveries of new ways of harnessing the sun, or greatly reducing the cost of already identified approaches. Solar energy offers us an outstanding example of a technological development that might be of enormous importance; however, severe constraints, none of them scientific or technological, greatly influence its rate of development.

Two applications of direct solar radiation are most often cited. In one, water circulates from roof panels, where it is warmed directly by the sun's rays, to a storage unit from which the heated water is taken for use. In the other application, the solar radiation falls on specially constructed metal photo-cells that convert the received radiation to electricity. The generated electric power can be used at that location or fed into the electric power grid serving an area. In both approaches, cost reduction and higher efficiency are needed and such progress depends on innovative processes and materials.

It is possible that sunlight may be used in the future for direct production of fuel in quite novel ways. For instance, in experiments being conducted by Dr. Harry Gray of the California Institute of Technology, a catalyst is dissolved in water which is then exposed to sunlight. The radiation and the catalyst act on the water to split its molecules, creating hydrogen which can be burned as fuel.* Efficiencies reached in these experiments have surpassed the efficiencies of known biological processes for producing energy with the aid of sunlight.

But there is much more to solar energy, including energy from the oceans, as already discussed, and from winds. Also to be included is the growing of trees and harvesting of them for fuel. The same applies to biomass in general, like raising grain to be converted to alcohol, this perhaps mixed with gasoline to form gasohol as a fuel for internal combustion engines. The moment we mention agriculture in relation to solar energy, it is important to call attention to how little we know as yet about the science underlying the process by which the sun's radiant energy creates fuel-rich molecular structures in those biological machines we call plants. However, scientists have uncovered knowledge in recent years that suggests it should be possible for us to improve the process greatly. With more research and invention, plants—whether improved forms of grains, trees, sugar cane, or some entirely new creations—might be caused to provide much more energy output for the same investment or input in land, water, chemicals, energy and mechanical equipment, and labor. We said earlier that our understanding of every aspect of food and nutrition is as yet meager. This is also true as to the agricultural approach to transforming the sun's energy to liquid and gaseous fuels and electricity.

An intriguing concept is to foster vegetation that yields a high percentage of hydrocarbons. Milkweeds and trees that produce natural rubber are examples. Experimental indications and theories based upon recent discoveries in biology suggest that further R&D may make possible the annual production per acre of many barrels of oil equivalent. Technological innovation is needed here because the best land is already being used to produce food, or wood that leads to building materials and paper products. Moreover, to plan a significant contribution to the energy supply problem it must be recognized how enormous the present energy demand really is. If the total harvests of all United States corn crops were fermented to alcohol, this would provide only a tenth of current gasoline requirements. However, genetic improvement and better management of plantings, from choice of species for a given area to techniques of cultivation, may make possible the utilization of land not currently employed.

*Hydrogen is in some ways the ideal fuel combining fundamentally unlimited supply, water being plentiful, and equally fundamental freedom from pollution, because when burned (combined with oxygen) it forms water.

As an illustration, Nobel Laureate Melvin Calvin has called attention to a tree in Brazil that produces as sap virtually pure diesel fuel. It has actually been tried successfully in the tank of a diesel vehicle. It is estimated that 25 barrels per acre per year could be obtained. Calvin has also expressed confidence in the plant Euphorbia, which yields hydrocarbons lending themselves to naphtha production at less than present prices.

Another potential energy source from biomass, but one requiring substantial additional R&D to bring to fruition, is the common cattail. This water plant is an efficient converter of solar energy and it is found on thousands of wetlands around the nation. The idea being proposed is that the cattails be compressed into fuel pellets or their starch removed and converted to alcohol. Since cattails grow naturally and best in the wetlands, their use for energy would not compete for land against other crops, and there is no water supply problem. Moreover—and this seems the ultimate in this "too good to be true" recital—cattails actually soak up some pollutants as primary nutrients. Cattail farms could even be created near sewage treatment plants where they would remove certain troublesome effluents. According to one estimate, the United States has over 100,000 square miles of real estate where cattails could be grown. However, if there is a potential for this crop as a partial answer to the nation's energy supply, it will be realized only if R&D is pushed.

There are other approaches to biomass energy conversion that await more R&D. Instead of burning wood to obtain energy, it can be converted, by further scientific effort, to valuable chemicals and fuels by using microorganisms. Fermentation microbes have already been used in producing industrial alcohol, but the specific advancement of the underlying scientific knowledge can provide other options.

One more illustration of what more science and advanced technology could offer us is aquaculture, the growing of food in water. This field of endeavor has great promise but so far has received little R&D backing. Particularly interesting is to create fish farms with unusually high outputs by raising the fish in the heated waters near conventional power plants. Over half the generated energy of present power plants is lost in waste heat, either directly up the stack in hot gases or in the cooling systems. It has been estimated that half of the fish to be eaten in the future could come from waste-heat aquaculture systems.

Both ocean and solar technology advances involve government funding, controls, arrangements, or integrated planning. Free enterprise, acting in response to market opportunities, is not going to have the option to determine what happens in these fields as though government did not exist. Even if the government remained aloof regarding the opportunities and problems in ocean and solar developments and made no attempt to interpose itself into the decision-making process, the private sector would be cautious and generally less than fully interested because these fields are mainly too long-term and speculative in nature. Private investment at risk is not consistent with situations in which government responsibilities as to basic resources (such as ocean waters) are in the end unavoidable, or where proprietary know-how developed at private expense is difficult to protect from competitors. Ocean and solar technologies are very likely to require a combination of government and private involvement. Such teaming in America usually presents difficulties.

In enumerating areas of research and development that illustrate the many facets of society's interaction with advancing technology we must not overlook controlled thermonuclear fusion of atomic nuclei. Fusion has the potential of providing us with energy that is plentiful, safe, and cheap. This possibility derives first from the inexhaustible supply of the basic raw material (fuel), isotopes of hydrogen, found in ordinary sea water. That fusion of atoms is possible is not open to question, since it actually takes place in hydrogen bomb explosions. But it will take considerably more R&D before it is learned how to control the process of fusion so as to make it a practical steady source of electrical energy. Nuclear fusion is in one sense the ultimate in solar energy production because it is the same source of sustained energy as is believed to power the sun. In one basic fusion reaction a plasma of hydrogen gas is made so hot that its individual nuclei fuse together to produce helium and release energy. Fusing the nuclei of hydrogen, the lightest natural element, releases more energy than the splitting (or fission) of uranium, the heaviest, which has been producing commercial electricity for years amid growing controversy.

The fission of heavy atoms derived from uranium powers all present nuclear reactors. At the rate at which these nuclear reactors are operating and might be expected to be expanded in number, there is enough readily available uranium for perhaps half a century. The entry of the breeder reactor, which makes much better use of the uranium, could cause the uranium supply to be extended for a few thousand years. Fusion, on the other hand, is based on matter that will be available as long as the oceans are here.

Unlike fission, fusion does not lead to generation of radioactive waste. It does not produce materials that can be modified and switched to create nuclear bombs. To bring about the fusion reaction, enormously high temperatures are required, and the containment of the high-temperature plasma in which the thermonuclear reaction takes place is the heart of the technology. The phenomena involved are inherently safe. If something goes wrong the process simply halts, the previously high-temperature plasma virtually instantaneously becoming cool, harmless matter. No runaway or bomb effect is possible. In fact, it is so terribly difficult to get everything right, that is, to provide the technical conditions that make possible a continuous, energy-producing reaction, that the scientists have been unable to do so after 20 years of trying. However, the containing walls gradually can be expected to become somewhat radioactive through spurious bombardment, and from time to time something will have to be done about that condition in any eventual full-scale installation.

Hundreds of millions of dollars have already been spent to explore the underlying physics and engineering of controlled thermonuclear fusion. It is estimated that decades more will be required before all of the remaining physical science and engineering details are worked out in sufficient depth for this approach to energy supply to be ready for application. It may turn out that it never will prove feasible. The total costs in the R&D and start-up phases surely will end up in the range of tens of billions of dollars. When we have to speak of funding at such levels, time durations of decades, and major doubts as to practical feasibility, we are beyond the boundaries of areas suitable for speculative investment by private entities. Energy from fusion, if it comes, will have to be by way of the government, not free enterprise. Yet, at some appropriate future time, if all goes well and stays well, the private sector will have to be brought into the act. How, in relation to government, is not now clear.

From the phenomena that light the sun, let us shift to another down-to-earth example. Oil production from the land in the United States can be increased either by discovering new deposits or by using advanced technology to extract more of the oil at the bottom of thousands of existing, already drilled wells. Free-flowing gushing does not characterize all wells, particularly after much of the initial, low-viscosity oil at high pressure has been tapped. When the oil reserves of the United States are estimated, more than half of the oil is not counted because it is assumed it will be left below permanently. So-called secondary and tertiary recovery techniques (which means the use of submergible pumps and the injection of water, steam, or chemicals) can provide enough oil to add years, perhaps even a decade, to the period in which domestic oil can supply an appreciable fraction of our total needs.

The costs of doing this are not small, but they would bring the approximate total cost of domestic supply only up to the prices presently being paid for OPEC oil. This recovery technology is not yet at the limit of its potential development and further technological advances could trim costs and further increase the available oil. Since added recovery systems would simply become part of the present infrastructure of the nation's existing wells, no major new environmental factor would be introduced and the process itself would be low in pollution consequences. Accordingly, regardless of what the nation might accomplish in energy conservation and in the development of alternatives to petroleum, it would appear that failure to apply and further advance this recovery technology can only worsen our energy problems. The United States certainly leads in the technology, but existing price controls on old oil provide no incentive to make investments in increased flow. With the government's rules as they are, if you own an old well, you ought to leave the oil in the ground for tomorrow. (That's probably what you are doing.)

A somewhat more exotic example is found in the potential of another form of energy extractable from the earth: geothermal energy, the natural heat of the earth. In various parts of the globe, this heat is available at a reachable distance from the surface, usually in watery fluids and steams that are in contact with hot, deep rock formations. The biggest bottleneck to progress here is government red tape, particularly in connection with assignment of land and the working out of suitable environmental regulations.

So far, we have introduced areas of research and development that affect human surroundings and the creation or capture and utilization of physical resources. Let us shift abruptly to look at the rewards on the horizon in medicine—the prevention, diagnosis, and treatment of the ailments of the human body. Because of recent discoveries by scientists in basic living cell behavior and other phenomena in the field of microbiology, we can expect rapid progress in very fundamental medical science.

One example is the use of recombinant DNA, a name for taking genes from one organism and splicing them into the genes of another. Through this technique microbes can be used to do such things as manufacture antivirus compounds, synthesize vaccines, make human insulin, or break down industrial waste. Millions of diabetics depend on insulin made from the pancrea of cows and pigs. If the product is created instead by bacteria, carefully and ingeniously implanted with the human gene, then the resulting matter will be identical to human insulin. Rejection by the body as a foreign element will not then take place. A billion-dollar-a-year DNA industry is not out of the question in a decade as initial work in pharmaceuticals expands into commercial, chemical, food, and fuel provinces.

But such employment of bacteria on a mass scale is in some respects equivalent to producing new hybrid microbes that could themselves be a health hazard if not under rigid control. Some scientists have warned that DNA manipulation activities might inadvertently create new organisms that could creep out of the laboratories or the factories and cause major epidemics. Others say that while some experiments are so dangerous they should be banned, other DNA effort based on the standard bacteria used in most activities are safe because those organisms are too weak to survive outside the laboratory or to harm human beings. Who is right? The National Institutes of Health recently set up an advisory committee to examine the dangers and, as the committee phrased recommendations, they arrived at a split vote (10 to 4 with one abstention) on relaxing present strict guidelines. Thus, as is typical with scientific and technological advances, if the R&D is carried on, we are required to include a parallel attempt to understand risks and then innovate so as to minimize them while maximizing potential gains. The U.S. government needs to regulate DNA work, but if the restrictions are too severe, the effort simply will move to Europe where the climate is less restrictive. This has already begun to happen. Seeing the broad applications of genetic engineering, European governments have been actively promoting it. European scientists are extremely competent in this field. Our leadership in genetic engineering thus is not at all guaranteed, even if American funds do not move abroad.

Another example of the potential of applying the latest genetic science is in the production of human interferon by bacterial factories. Interferon is a natural substance that fights viruses and tumors. It is produced by almost all cells of the human body by their use of instructions contained in a specific gene of their DNA. After extracting the appropriate genetic substance from human white blood cells which were producing interferon, scientists have processed it and spliced it into the genes of laboratory bacteria. The bacteria then proceeded to make clones or copies that contained the human interferon. Through this recombinant DNA approach it is possible that future developments will yield a low-priced source of pure interferon with powers to fight many diseases, from common colds to cancers. In tests, doctors have found that human-produced interferon will combat hepatitis, herpes virus, eye infection, and certain infections of the newborn.

We could cite many more possibilities in medicine, based on further R&D in microbiology. However, it will serve our purposes best to focus next on examples of technological advances of great potential in medical care, as distinct from further scientific discoveries.

As one example, microsurgery, a name for any surgical procedures done with the aid of microscopes, has already benefited millions of patients. Newer techniques, made possible through additional R&D, could exert a profound impact on virtually every area of surgery. Ophthalmologists have pioneered the use of the microscope in surgery on the front portion of the eye to remove cataracts and to transplant corneas. The added possibilities include preventing certain kinds of strokes, coping with some types of cancer, aiding people injured in accidents, correcting birth defects, restoring vision and hearing, and correcting damage from inept surgery.

As another illustration, further R&D may yield in a reasonably short time a blood substitute made up of biologically inert chemicals that can perform the function of red blood cells in carrying oxygen. It appears likely that a significant share of the estimated 10 million units of whole blood used yearly in the United States will become replaceable. Unlike human blood, the fluorocarbon emulsions making up the synthetic blood have a shelf life of several years even at room temperatures (in comparison with whole blood which has a shelf life of weeks under refrigeration). The blood substitutes, if all works out, can be used on anyone regardless of blood type, and will not carry infections such as hepatitis.

The secret to the life-sustaining process of this substitute blood is that perfluorochemicals—hydrocarbons in which the hydrogen atoms have been replaced by fluorine atoms—can dissolve almost 60 percent oxygen by volume while whole blood can take up only 20 percent and blood plasma only 3 percent. Unfortunately, some of these new chemicals, while carrying oxygen so advantageously, tend to concentrate in certain organs of the body and must be cleared out. Frontier R&D is now being done to speed the clearing process. Japan is a leader in this effort. Japanese law permits routine use on general patients as soon as a new substance has been tried without apparent damage on 150 patients. Our rules are more severe and less specific; thus it is likely Japan will lead us in this area.

New measurement techniques depending on sensitive electronics and precision x-ray instruments backed up by computers may revolutionize medical diagnostic procedures. For instance, with more R&D, the body's skin and inner physical makeup could in the future be scannable with a degree of precision and microscopic detail that could reveal onsets of deviations from normal which today's physicians have no way of spotting. Tiny clots, stones, pollops, and other abnormalities could become discernible.

The so-called CAT scanner, a computerized x-ray machine that provides a three-dimensional view of parts of the body, has been regarded as so fundamental an advance in medical diagnostic techniques that its orginiators have shared a Nobel award. Yet the CAT scanner has become a controversial medical tool. It costs up to $1 million per installation and has been used as a symbol of a political medical issue, namely, whether we can afford technological advance in medicine. All diagnostic centers, even clinics and hospitals in small towns, are expected by their patients to possess a CAT. Otherwise those being diagnosed believe they are being discriminated against and disadvantaged, that is, denied the full health protection to which they are entitled. The example shows that our system for setting priorities is not perfect. We do not have ready means for setting health delivery standards, for balancing dollars against health improvement. We must select which way to spend available funds so as to buy the most in health benefits, and determine how much of all this we want the government to be deciding for us.

Another example worth citing is ultrasound, the use of sound waves to produce images of internal arteries, glands, tumors, fetuses, or organs at work. The method has only recently been introduced, but in time it may not only extend diagnostic medicine but be a safer substitute for some x-ray techniques. However, a safety hazard may exist in a new dimension. Without time and effort, the potential deleterious effects on the body of the inaudible, high-pitched acoustic waves will not be discovered and assessed and the limits of their use cannot be set. This new technology is also an example of growing costs and possible economic imbalances in medical care as we strive to bring beneficial new developments into use.

A particularly important potential area of medical technology results from a combination of two advances: first, a better understanding of how signals are transmitted from our senses to our brains, and second, the ability to synthesize microcircuitry to replace damaged portions of the communications system of the body. Experiments are commencing on restoring vision to the blind and hearing to the deaf. Electronic stimulation is provided to the brain of a blind person by arranging that the scene be viewed by a television camera which in turn is connected to electrodes implanted in the would-be viewer's brain. In the future, a highly miniaturized television camera might be placed in the eyes of a human being who has lost the power of vision and connected to several hundred tiny electrodes on the brain. A similar concept applies to the stimulating of auditory nerve fibers electrically.

As still another example in medicine, many drugs that would otherwise be useful to treat specific diseases cannot be prescribed because of serious toxic side effects. When a drug is taken through the digestive system, the portion of the dose that gets to the right place for beneficial results is sometimes inadequate, while the amount that interferes with the rest of the body can be very high. New techniques are in development that provide controlled time release. Initial experiments suggest that implantable devices can be made practical to deliver effective, essential matter to the body on a controlled time schedule and with focused concentrations at specific points, a function that oral or intravenous drug administration cannot achieve. Examples include a reservoir implanted under the scalp to provide access to the otherwise essentially inaccessible cerebrospinal fluid. For this, a bulbous unit is located close to the scalp and under the skin where it can be refilled with a hypodermic needle when necessary. The drug drains from the reservoir into a ventricle of the brain. Another example is an implanted pellet which releases its concentrated drugs into the subcutaneous tissue through a process of erosion. Still another example is an implanted miniaturized pump driven by a chemical propellant that maintains a constant pressure to deliver a steady supply of a liquid drug to a vein or an artery.

Total medical care is desired by the American citizen but complete attention from birth to death cannot be provided for all by present methods and facilities, or a mere extrapolation of them. The expenses are too high and the required resources are greater than can be made available. There is no end to the additional health services that can be conceived and developed, and trying to do too much drives costs up and hurts quality. To arrive at the best balance we need more efficient and economical systems of preventive medicine, medical testing, and data collection and processing, and much creative and analytical work in design of hospitals, clinics, and other health care facilities. Such effort requires substantial investment, but in the end costs could be reduced substantially. An example is in the application of computers to the medical care area.

Computers can be tied to numerous measuring devices to assist in obtaining, recording, and processing diagnostic data about a patient, in the running of the hospital, or in the monitoring of patients. Basically, the computer can decrease the expense of handling the information that is key to good medical practice. For instance, the computer can help tend patients in intensive care, instantly responding to critical signs and sounding alarms. More broadly, if adequate R&D is carried out and the results are applied in practical implementations, the practicing physician of the future will be able to take a patient's information—background, laboratory tests, symptoms, and complaints—and feed it into a diagnostic data network by pushing buttons on a special electronic terminal wired into the systems's center. At this center a constantly updated compilation of the results of treatments of diseases will be kept. This repository of facts will disclose statistical correlations of cause and effect, ailment and cure, drug and symptom.

One aspect of this kind of medical information network, the straightforward part, has to do with the transmission, storing, and routine processing of information. This part is no more complicated in principle than the electronic reservations system already used by the airlines. We are all accustomed to approaching a ticket counter, describing where we want to go, and seeing the clerk address an electronic terminal and read off some information, to tell us whether we can have the reservation—all in a few seconds. The electronic input device enables the clerk to send signals over a telephone line to a computer which has records of all the flights and the reservations already committed. Rules are stored in the system for handling the signals so that the inquiry is automatically paired with the schedules and the available open seats. A conclusion is reached and the computer sends an answer back over the telephone line to be displayed to the operator.

The hardware needed for a medical information center is very similar and is in part already available. The much more interesting question is what to make of all the data so that medicine can be practiced with higher skill and less cost. The design of the system will have to be based on statistical analysis and creative medical and engineering effort to develop ways to use the information beneficially. For instance, inquiring physicians might see on the electronic display device a statement of what the odds appear to be for various diagnoses and treatments, even perhaps the specification of a range of strengths of medication, all this based on the reported accumulated experience on previous similar cases. Of course, the physicians would have the obligation to report their own work to the system. As time goes on, in a continuous updating, the data system would increase its knowledge of the total experience as to effectiveness of different medical approaches. It would be as if a physician could consult instantaneously and simultaneously with thousands of colleagues who have dealt with like problems. Compared with what is possible in the future, physicians today are not in a position to share their experiences well. They do so only to a limited extent and after years of delay. The technological society ahead will see a tremendous step-up in the degree of communication and in the amount of information available.

The use of the computer to improve medical care and lower expenses for quality results also will broaden medical practice and bring into existence new professional and supporting endeavors, technicians, apparatus producers and designers, maintenance staffs, record keepers, electronic machine operators, and others. The government's duties in medical care will also broaden. Medicine already has become a field of endeavor with ever larger government involvement, from the control of prescription drugs to the dispensing of funds for medical care. The entry into medicine of computer networks will bring further need for government in regulation and setting of standards.

Consider next the practice of law, as it might be affected by computer information systems. In the future, the office of a typical practicing attorney might have a convenient connection to a central electronic file of cases, statutes, regulations, procedures, and commentaries. The attorney will be able to query this repository by operating an intelligent electronic terminal, looking a little like a typewriter and TV set combined. Almost immediately after the inquiry, there will be displayed on the viewing screen pertinent information available on a particular question.

Such an electronic law information network would serve the lawmaking bodies of government as well, not only in the fashion just described, but in additional important ways. For instance, a legal information system cannot be designed without giving the laws and regulations of the land a good going-over to pinpoint their essences, conflicts, and shortcomings. Since an electronic information system of the future can accommodate masses of data on the results of enforcement of existing laws and regulations, and on violations and litigation among individuals, corporations, and governments, such comprehensive statistical data should help to disclose where existing laws and regulations are ineffective and where different and superior rules are needed to keep everything clear and fair. Obviously, the entry of this kind of electronic information system into law practice, law making, and law enforcement requires the cooperation of government, which has unavoidable responsibilities, and the private sector, which has indispensable expertise.

We have in this chapter listed examples of areas in which it is likely that advancing technology can provide benefits so great we would not want to forego them. For economies and social advantages, further advances in these areas should be sought. We also were able to perceive some associated negatives, a few easily handled to subdue or limit them, others perhaps more formidable. Nothing about these potential handicaps suggested the bad must necessarily outweigh the good. However, the tradeoffs of potential positives and negatives should be researched to provide a basis for decisions as to the magnitude and nature of sensible implementations.

Tentatively, we could say it would be advantageous to the United States if we were to lead in exploiting these advances. It would portend for us a better life and more products for export, yielding us the means to buy the contributions of others. For none of these examples are we compelled immediately to assume it is absolutely vital the U.S. be No. 1. Presumably, if we could be ahead in some of these areas of technology, we could countenance other nations' being more effective in the rest, assuming a reasonably free trade world. If we lag always in every technological advance we shall suffer in living standards while also failing to make our proper contribution to the world society.

Finally, in every example we cited, we could see some non-technological factors controlling the rate of advance; these lie in the social–political–economic arena. We are ready now to consider further these non-technological factors and their interrelationship with technological advance. We can best do this by facing directly the question of America's overall status in science and technology.

3

a slip that's showing

Is the United States losing its position in developing and applying new technology? We remind ourselves that slipping badly here could be severely penalizing to our living standards, our ability to fight inflation and unemployment, and our freedom to enjoy a preferred way of life in a highly competitive and dangerous world. Let us examine the available evidence, beginning with a look at a measure commonly cited when concern is expressed: the productivity of American industry. For more than a century the United States steadily improved its output of goods and services per worker. We led the world in inventing new processes and building more efficient facilities and equipment. We left other nations far behind and yet our superior productivity kept improving year after year. All this was true until the last decade.

By 1980, the United States had managed to become last among the world's industrialized nations in annual rate of productivity growth. Our yearly productivity increases are hovering at zero or have actually gone negative,* while other nations—France, Italy, the United Kingdom, Japan, and Germany, particularly the last two—are attaining substantial positive productivity rises. From the mid-1940s until the mid-1960s, American productivity increased over 3 percent per year. Between the mid-1960s and the mid-1970s this measure dropped to just over 2 percent. For the last half of the 1970s the yearly rise sank to an average of 1 percent. A negative figure is on the report card as we start the 1980s. All in all, from 1950 to the present time, productivity in Japan rose at a rate four times the U.S. average.

---

*A negative annual rise means that instead of becoming smarter or more productive than the previous year in what we do—presumably because of more experience or inventions or the installation of newer and more efficient methods and machinery—we became less able and effective, as judged by how much we turned out.

In the mid-1960s it took some 25 worker-hours to produce a ton of steel in a Japanese mill while the labor required in the United States was half that amount. By the mid-1970s the U.S. figure improved to 10 worker-hours, but the Japanese meanwhile brought theirs down to 9. Japanese steelmakers have facilities much more efficient than ours — big, well laid-out plants using the latest technology, large blast furnaces, more extensive use of continuous casting, better integrated operation, and expanded computer control. In the past 20 years, Japan has built several completely modern plants; the United States has built one. The energy consumed to make one ton of steel in Japan is now about a third lower than that in the United States.

So far we have been speaking of productivity increases, not the actual magnitude of productivity. Despite the fact that the enhancing of our productivity has stopped, our output per employee was still the world's highest as we completed the 1970s. However, if the trend continues for another decade, West Germany and France will have passed us and Japan will turn out more per worker in four more years. Relative productivity for nations is almost synonymous with relative level of real income and material well-being. It also measures important elements of relative economic strength and the capacity of a country to assign resources to areas of its choosing such as energy, transportation, health care, and military preparedness. The Joint Economic Committee of Congress has warned that the average American's standard of living is likely to be drastically reduced in the 1980s unless productivity growth in the United States is resumed strongly and quickly. According to the Committee's study, faltering productivity means a stagnating economy with fewer Americans able to afford a decent home, the slow growth particularly handicapping the disadvantaged minorities.

Some mistakenly believe the loss of productivity gains is a cause of our rising inflation. The cause of inflation is not low productivity. Inflation has other roots, notably in overly rapid rates of money supply growth. Nevertheless, productivity is pertinent to success in handling inflation. If productivity grows sufficiently, business finds it possible to increase real wages without increasing prices and still have funds left over for expansion and further innovation. This has been the typical American pattern for many decades. Inflation goes with lagging productivity growth when the government acts as if productivity were rising when it isn't, and pumps up the money supply. This usually happens in response to political forces, as when an attempt is made to heat up the economy and reach unrealistic targets for unemployment levels. If at the same time wage expectations are met even if productivity gains lag, then we get undesired increases in both the money supply and prices along with the desired increases in wages. During the first half of the 1970s compensation per hour rose about 8 percent per year while productivity increased about 1½ percent annually. Wages and fringe benefits constitute the major components in the costs of all goods and services produced. To the extent that labor cost rises faster than productivity, the difference must find its way into prices. Thus, in slowing our rate of productivity advance we have not helped ourselves in combating inflation. To appreciate the effect of the lag in recent years, take 1978 as an example. In that year, a typical American family had $5000 less real income than it would have had if the growth in productivity had remained at historic levels. The total loss in goods and services available to the citizens of America because we failed to maintain the productivity increases of the mid-1940s was worth $300 billion in 1978 alone, more than twice the entire defense budget for that year plus the price of all the automobiles produced.

Our deteriorating trade balance is as much a cause for alarm as what is happening to our productivity. In 1978 the United States trade deficit reached a record figure of about $30 billion. It is often stated that this was the result of our having had to import oil at prices much greater than in previous years. But in 1978 the U.S. imported only 25 percent of the total energy it consumed, while Japan imported 95 percent and Germany 60 percent of their respective requirements. In the same year, Japan enjoyed a $63 billion and West Germany a $49 billion surplus in manufactures, a category in which the U.S. experienced a deficit of about $10 billion. The problem for us is not alone that we had to import oil. Those imports hurt, but a worse problem is that we appear no longer as effective at converting materials and energy into manufactured goods as certain other countries. They have been taking from us much of our domestic and overseas markets. If they were not superior to us in manufactured goods on the average, their more adverse situation in energy imports would have caused them to register serious trade deficits. Instead, it is we who merited them.*

We are now taking in manufactures at two and a half times the rate of oil. Between the early 1960s and the late 1970s our imports of manufactured goods increased by a factor of 20! Our biggest loss of trade position has been in that area labeled "mechanical systems,"

---

*As this book goes to press, our overall trade balance seems to have improved somewhat, but not in the critical area of manufactures.

which represents about 75 percent of our trade. It includes four
categories: heavy machines, light machines, cars and trucks, and
airplanes and space vehicles. Only in the last category do we remain
relatively strong, with exports around $8 billion and essentially no
imports. This one favorable category may not remain so, however,
because the Japanese and Europeans have plans to invade this
market strongly during the 1980s.

For two decades the Europeans and the Japanese have been
dependent on the United States for space satellite boostings. When
the U.S. recently did the French the favor of launching their communications satellite we were able to exact the condition that it not carry
commercial traffic to compete with U.S. satellites. But now western
Europeans are challenging the U.S. in the launching business. Having
produced a booster rocket of their own, they have already taken
some business originally planned for NASA's new shuttle spacecraft.
They appear determined to win in forthcoming competitions for world
satellite programs, including one system for Indonesia and another for
a group of Arab states. The Scandinavian countries, with their sparse
populations, great distances, and mountain barriers, are planning a
local TV satellite system for the mid-1980s, and Germany and France
probably will soon develop Europe's first broadcast satellite system
relaying signals directly to European home TV sets. It seems western
Europeans are prepared to buy their way into the space race if
necessary, by bidding low for the first several years. Considering
everything, the chances are that the United States may not continue to
monopolize the non-communist world's space activities.

We are now importing almost $12 billion more automobiles and
trucks than we export. In entertainment electronics, we are off by $4
billion, in heavy machinery by over $6 billion. In the light machinery
category represented by, or used to manufacture, most consumer
goods—typewriters, sewing machines, furniture, cameras—we are
showing a deficit of over $12 billion. The total deficit in U.S. trade in
machinery and all of the manufactured products based on that
machinery is some $35 billion. This means we are losing the jobs and
the accompanying tax base associated with these essentially clean
industries and aggravating our payments imbalances and the problem of adjusting to the economic impact of importing so much oil.

It is especially detrimental to America's economic strength to lose
superiority in technologically intensive industries. Those areas of endeavor greatly dependent on advanced technology have performed
much better in recent decades than the sectors of the economy that
are of low technology content or are essentially non-technological.

The real economic growth rates have been higher, so new job opportunities have been greater. Productivity in these industries has also increased more than in other areas with the result that the price inflation rate has been lower.

Four ways to improve productivity are often mentioned: technological innovation; heightened capital investment; better training and education, and higher motivation of workers; improved government–business relationships. We appear to be headed in the wrong direction on all of these fronts. Studies have indicated that about half the past increase in productivity in the United States can be attributed to technological change, that is, to a combination of scientific and engineering advance that yielded improvements in the way we produce goods and in the know-how of management. No precise way exists to measure the impact of technological advance on overall real economic growth, but as with productivity, technological innovation again is a strong factor. Those who have studied this relationship intensively have estimated that technology advance has given rise to between a quarter to a half of the entire growth of the American economy in the past several decades. Around three-fourths of our manufactured goods exports are usually labeled technologically intensive and half of our manufactured imports are in this category. If agricultural products are counted as technologically intensive — with our mechanical equipment, fertilizers, insecticides, herbicides, and preservatives, this seems a realistic categorization — then the impact of our technology status on our exports is a penetrating one.

In toto, new technology is highly influential in general economic growth, productivity increases, and trade balances. Unfortunately, the United States spent less than 2 percent of its GNP (gross national product) in the 1970s on non-defense R&D while Japan's and West Germany's ratios were substantially higher. It has been reported that for 1980 U.S. manufacturers expect R&D to account for less than 50 percent of their capital spending whereas in the mid-1960s this figure was close to 80 percent. In a not surprising correlation with reduced technological innovation support, the real growth of the U.S. economy in the 1970s was substantially less than in the 1950s and the 1960s. In the 10-year period from 1966 to 1976, the real growth rate for Japan was 9.4 percent, for France 5.9, for West Germany 4.7. Only the United Kingdom at 2.7 percent was below the U.S. figure of 3.4 percent. During that same decade the value of the U.S. dollar relative to the deutschemark and the yen declined dramatically (down to almost half in the case of the deutschemark and down to two-thirds of the beginning ratio in the case of the yen). Many blame this on the high cost of energy and point to the 1973 OPEC oil embargo as the turning point. Actually, the currency exchange ratio falloff started with vigor before the completion of the 1960s.

Our share of world exports has declined some 25 percent during the period from the early 1960s to the late 1970s, while West Germany and Japan increased their fractions. During this same 15- to 20-year time frame the percentage of scientists and engineers engaged in R&D in the population increased by over 70 percent in West Germany and over 100 percent in Japan. Even the United Kingdom showed an almost 40 percent increase in this population ratio. In the U.S. we found ourselves at the end of the period with fewer engineers and scientists than at the beginning.

Engineering manpower at the doctorate level is important in high technology innovation and development. In the 1970s the annual rate of Ph.D. graduates in engineering in the United States fell by over 25 percent and an increasing fraction of these new doctorate holders were foreigners, their number averaging over a third. There are neither enough faculty in the universities nor U.S. citizen-students pursuing engineering doctorates today to create a near-term reversal in this trend.

Unlike the situation in some competitive nations, a close relationship is lacking between our universities and the manufacturing industries in the field of production technology. Only about a dozen universities out of several hundred have strong departments in manufacturing engineering, most with only light relationships with industry. This is not enough to lay a proper base for leadership in manufacturing. For example, in metalworking, in many aspects of which Japan and Germany are besting the United States, greater use of computerized control and other advanced techniques are key to raising productivity. American graduates in engineering are rarely given a background to permit them to merge production techniques and computers effectively. Moreover, only the large corporations typically have on their staffs engineers thoroughly able to adapt computerized systems to machine control, i.e., capable of synthesizing and analyzing novel operating methods, estimating the costs versus benefits of new means for making products, and designing the complex computer–man–machine partnerships required to increase productivity and decrease costs.

Total applied research and development in United States industry, whether through government sponsorship (as in military, space, and nuclear energy) or private investment (to develop new products and manufacturing techniques) has not kept pace with inflation over the last decade. Industrial R&D as a fraction of GNP has fallen from a figure over 3 percent to one just over 2 percent. During this same period the total R&D budgets, again expressed as a fraction of the nation's overall domestic product, rose in France and Germany. If we look carefully at the makeup of our R&D programs as compared with those of other nations, we discover that the budgets for R&D in U.S. industry that is entirely company-financed have kept up well. The reason for the falloff in total R&D in the U.S. has been the cutbacks in defense, space, and nuclear R&D, all sponsored by the government. Apparently, in view of the negative results achieved in recent years by the U.S. in productivity and trade balances, our industrial R&D is not focused well on those aspects of technology which have most to do with productivity and exports. Other nations have done a better matching job.

We have to recognize that government-supported R&D is not always pertinent to productivity increases. Federal funding for medical research, weapon systems, and NASA space projects has less direct effect on productivity than private industry spending on technological advance, even though government programs occasionally provide important offshoots. It is also well to note that in trying to gauge a possible deterioration in U.S. R&D endeavors, it is certainly not sufficient to measure R&D budgets against GNP. Such a comparison might be pertinent to the manufacturing portion of GNP, but the service sector of the nation, where relatively little R&D is undertaken, has constituted a rising fraction of the GNP over the last 10 or 20 years.

Putting aside ratios for the moment, the United States still has the highest total expenditures on R&D and a larger fraction of our population can be characterized as scientists and engineers as compared with other nations. So while there is reason to be concerned about the trends, we should not rush to negative conclusions as to our present strength and potential for future improvement. Still, it is hard to feel comfortable. Of the federal government's huge R&D programs, practically none is directly definable as advancing the engineering of machinery, the area that figures so greatly in our bad trade balances. Government-sponsored research to improve general engineering techniques (as distinct from defense, space, nuclear, and other specific mission endeavors) is less than half a billion dollars a year. Of this only some $25 million is allocated to research in the basics of mechanical engineering even though this category is fundamental to machinery, our severest export–import problem. Even after we add industry-sponsored R&D to the government's program, the total of all

R&D funds applied to the machinery area, where we are losing out badly to other nations, is still only 5 or 6 percent of total American R&D.

To illustrate further the relationship of R&D to improved manufacturing productivity, it will be helpful to quote from a recent advertisement of the Western Electric Company:

> In the last ten years, the costs of labor and raw materials that go into the telephone have skyrocketed. And yet, the price of the finished product has risen only 18 percent. The explanation is really quite simple. During the same period of time, the telephone has been redesigned literally dozens of times. Western Electric engineers kept discovering ways to make it a little more efficiently. So while our materials and labor costs were going up, we've been able to hold our manufacturing costs down. Not a bad way to cope with inflation. At Western Electric, cost reduction and other improvements in our productivity don't just happen. They're the result of a systematic and formalized program which has existed for years. One of the key elements in this program is our Engineering Research Center. Western Electric is one of the few corporations that has such a facility. Unlike most research and development centers, its primary purpose is manufacturing research. In 1977, the net effect of our cost reduction program was a savings of over $200,000,000. With cost reductions like these, no wonder our rate of productivity improvement is well ahead of the overall U.S. rate.

Equally, we can quote excerpts from a recent IBM advertisement:

> A set of computations that cost $1.26 on an IBM computer in 1952 costs only $7/_{10}$ths of a cent today. That's because IBM scientists and engineers have put their imagination to work to create and improve information technology. The computing power of a machine which filled a large room 25 years ago, for example, is contained in circuits that now can be held in your hand. And the computation speeds are over a thousand times faster. With every innovative advance, and every resulting reduction in computing costs, the advantages of information technology become available to more and more people. Advancing technology increases productivity. And greater productivity can indeed help bring costs down to size.

These examples might suggest that all the U.S. needs to do to increase productivity is increase industry R&D. Even if this were true, it is easier said than done, as we shall see shortly. But it is not true. The focus of R&D counts as much as the size of the budget. The trend of our R&D spending admittedly is down compared with important competitor nations but in the past few years U.S. overall funding has exceeded such individual nations as Germany and Japan while their productivity gains have outstripped ours. Apparently, as we noted earlier, their R&D has gone more directly to productivity. Ours has included much more emphasis on military weapons developments and the meeting of government mandated safety, health and environmental requirements.

The economic advance of a nation depends greatly on science and engineering know-how but more specifically on the speed and manner in which knowledge is put to use. The size of the basic research program and the number of good inventions conceived are pertinent, but follow-up is needed to create products that influence life beneficially. Otherwise we merely uncover new techniques to leave them sitting naked. Advanced research explorations in the United Kingdom have been more outstanding than their selection of engineering projects. The Japanese have been very competent as fundamental science and technology advancers but they have been incredibly remarkable in innovative adaptations of their own and everyone else's breakthroughs. Many years ago, it was the low cost of their highly skilled labor that gave Japanese industry an advantage. Now the advantage is in how they employ their engineering effort. Consider, for instance, their approach to TV set design. Noticing that the cost of repair was becoming so high it was a major issue for the customer, they put their major effort on redesign for unprecedented reliability. They also curbed higher labor costs by an unusual degree of automated fabrication, assembly, and test. Japanese TV producers now have lower overall labor costs, lower warranty expenses, and decreased incremental costs on higher volume.

The British–French *Concorde*, their supersonic transport, is an example of misplaced or mistimed investment in advanced technology. The British and French have lost billions of dollars in the project and stand to continue losing more to keep it going. The project is simply not economic. There never was a market for tickets at a price that would make the plane profitable. Not a single plane was ever sold commercially — they all went to the two government-owned airlines. The goal should not be to create the most advanced technology but rather to put available resources into those product areas that underpin profitable domestic and world markets.

Competing with R&D in importance for nurturing productivity growth is capital investment. Unfortunately, we are also falling in rate of invested capital. It is not amazing that the productivity improvement lag in the United States has followed our capital investment dropoff.

Moreover, the composition of capital expenditures in the U.S. has been shifting markedly away from support of innovation in design of products and techniques for manufacturing them. Of the total dollars invested, an increasing fraction is now being earmarked for meeting environmental and occupational health and safety requirements. If we examine U.S. funds behind each worker and subtract that part of the investment made to comply with air and water pollution and occupational safety regulations, we find the amount has actually declined (expressed as a ratio to value added during the manufacturing operation under consideration) in the manufacturing sector of the U.S. economy since the middle 1960s. The capital went down from $258 per worker in 1967 to $220 in 1973. In the same period, this indicator rose from $298 to $693 in West Germany and from $191 to $324 in Japan. Overall, the U.S. invests only 15 percent* of its GNP; West Germany invests 22 percent and Japan invests 29 percent. Our failure to keep up in capital investment is certainly serious because it suggests an unwise deferring of the badly needed updating and replacement of existing facilities. If we continue to underinvest, we shall surely undermine our ability to produce with reasonable efficiency.

The nation as a whole has a low ratio of investment to consumption as compared with Japan, Germany, and other developed nations; this ranking unfortunately corresponds well with our lower productivity gains. To equal Japan's ratio of investment to GNP we would need to find $200 billion more to plow back annually. The United Kingdom, with the second lowest investment rate, also had the second lowest productivity increase. Of course, the U.S. difficulty in capital investment stems directly from the fact that the government taxes away funds from the private sector that might have been used for investment and innovative effort and uses them to cover the government's proportionately high expenditures.

In some of the underdeveloped countries, the standard of living is so low that to hurt it in the short term in order to fund investment that will later build up the economy and improve conditions is not a realistic possibility. Forced restriction on consumption for the masses of the people can create deepening poverty beyond a containable threshold. Then repression and loss of civil liberties under a severe, fascistic rule come about to counter the increasing possibilities of revolt. But this sad tradeoff of investment for growth versus the im-

---

*As this book is being written, the rate is falling drastically from this average figure for the 1970s.

mediate social condition of the people should not apply to a highly developed and relatively affluent technological society such as the United States. Here, if in strong majority we were agreed on it, we could forego some near-term consumption on the part of the majority, who are well above any poverty threshold, in order to increase investment and later elevate everyone's standard of living. If this were done equitably (for instance, to shield the poor from the blow) it later would improve productivity and the economy overall. This would then benefit everyone, including those who are below the income threshold goal considered appropriate for U.S. citizens. If capital formation were higher and consumption lower, the nation would have the funds to build urban rapid transit to cut energy use and move people more cheaply and cleanly than private cars do. With more capital to invest, we could modernize the railroads, which again use less energy and move most freight cheaper than trucks. We could build newer plants that operate with higher efficiency in major industries such as steel. We would certainly invent new products and new production techniques to increase productivity, decrease manufacturing costs, raise exports, and lower unemployment.

But, alas, there is more to recite suggesting a serious American technology slip. Patents awarded to Americans peaked in 1971 and the number issued annually has been dropping several percent per year. The fraction of United States patents granted each year to foreigners has doubled in the last decade. In the past five years non-U.S. citizens have won more than one third of all the patents issued by our government. Meanwhile, the share of patents held by Americans abroad has actually declined. Conceivably, the fact that foreigners are acquiring a larger fraction of U.S. patents than in earlier years might be because they are filing many minor improvement patents on our more basic, earlier inventions. The statistics do not show the value of individual patents, but only the quantity. The U.S. is virtually an open trade area for the rest of the world. Thus, with other nations catching up in technology, it has become in their interest to go after our market more assiduously and seek broader patent protection in this country than would have made sense for them previously. Meanwhile, another sober statistic: in the 1950s over 80 percent of the major inventions brought to market in the U.S. were developed here. By the late 1960s the figure had declined to just over 50 percent.

Both continuous casting and the basic oxygen process in the making of steel were refined in Japan. Disc brakes, radial tires, and rack and pinion steering appeared on European cars before they were introduced into the United States. British scientists invented both the hip replacement device that has helped so many victims of arthritis and the CAT scanner that has revolutionized medical diagnosis. The French pioneered the discovery of the phenothiazine drugs that have radically altered treatment of psychotic patients. Studies have been made seeking to identify the most significant technological advances in recent years and, whereas a decade or two ago the United States dominated the list, other nations have been first in the past decade in an increasing number of areas. These studies support the common-sense conclusion we are led to by merely observing the radios, hand-held calculators, digital watches, tape recorders, TV sets, hi-fi systems, cameras and, of course, motorcycles, automobiles, buses, and trucks that have entered the U.S. market from abroad and totally taken it over in many instances.

The evidence of an American technological lag extends into the security field. In the years since the first Soviet Sputnik, we have received steady reports of military technology advances by the Soviet Union. It seems grossly oversanguine now to conjecture that the United States still leads in all military technology — not while we are being told that today the Soviet Union has the largest intercontinental missiles carrying the largest numbers of the largest yield nuclear warheads. It does not make for confidence when we recall it is decades since the U.S. alone had the A-bomb while the Soviet Union was playing catch-up, or that twenty years ago we already had an ICBM capability in place and the Soviet Union then was only beginning to get its missiles into production.

The Soviet Union is spending roughly a third more on defense than we are.* While our defense budgets in real terms have remained essentially fixed for years, the Soviet Union has been maintaining a 4 to 5 percent real growth in expenditures for more than a decade. For the past decade or two the Russians have averaged about 13 percent of their GNP for defense, while U.S. defense spending has been around 5 percent of GNP. The R&D budgets that go with these figures lead to an estimate that the Soviets have spent $40 billion more than the U.S. in the past decade. Soviet capital investment backing up military R&D has exceeded ours by roughly 75 percent. There is a distinct Soviet production volume advantage in many categories of military equipment, typically 2 or 3 to 1.

But if there is conspicuous evidence that we are slipping in our

---

*In total R&D, the USSR commits 3.4 percent of their GNP, a roughly 50 percent greater ratio than that of the United States.

leadership in technology, there is much to cite suggesting that we are still ahead on points in the pure science olympics, even though we do not win every event. We have more Nobel laureates than the rest of the world combined. We are still strongest in solid-state technology (where we were first with the transistor), large-scale integrated circuits (complex electronic circuitry on tiny chips), digital computers, digital communications, and other aspects of information technology. Even though the Soviet Union holds the world record for the largest number of hours a human has spent in space, we have on the whole accomplished much more in putting useful apparatus into space reliably. We landed men on the moon and brought them back alive and healthy while the Soviet Union had great difficulty trying to land pieces of equipment there. Today our communications, earth observation, weather, and interplanetary research spacecraft surpass U.S.S.R. performances in these areas.

However, much of this impressive record is a description of the past. Basic scientific research, mostly sponsored by government and accomplished in our universities, has also suffered in the U.S. in recent years. After a damaging funding downturn in the late 1960s and early 1970s, the Ford Administration reversed the trend and Carter's has continued to improve the support.* However, after taking into account inflation† and the increasingly burdensome administrative requirements that now accompany research grants, U.S. basic research sponsorship has to be judged as not keeping pace. Such research, generally understood as a requisite planting of seeds for future technological advance and long-term improved economic strength, is increasing in other countries.

Turning from pure to applied research, the largest concentration of expertise in applied science and technology resides in the private sector of the U.S. Here one finds scientists and engineers with know-how in applying proven techniques as well as professional understanding of frontier science and the methods of advancing and matching it to specific requirements of the society. The industry has systems for arranging financial backing and experienced management to carry out successful implementations. Unfortunately, this pri-

---

*Thanks in substantial part to an excellent presidential science advisor, Dr. Frank Press, and a Secretary of Defense, Dr. Harold Brown, who as a scientist also understands the long-term importance of research.

†The inflation rate for R&D (facilities, laboratory instruments, technical manpower) is about twice the average price inflation for the economy as a whole.

vate sector has had less funds available in recent years for carrying on R&D and backing innovative technology than earlier. The most direct reasons for this are inflation and high taxes. The two go together and reinforce one another, and both stem from government policies and practices.

In computing earnings (the difference between revenues and all costs of doing business, including taxes), private industry must use the rules of the Internal Revenue Service (IRS) governing the important factor of depreciation, which covers the cost of maintaining and replacing facilities and equipment. The IRS requires that depreciation be figured on the basis of the historical price, that is, the price of the depreciating items when first bought. Because of inflation, replacement prices usually will have more than doubled during the life of the equipment. When industry reports its earnings, it is thus forced by U.S. tax law to exaggerate them, not being allowed to take into account the real costs of replacing its capital assets. Similarly, much of the reported earnings of American corporations include the unreal profits from selling off inventory at a higher price than that at which the inventory was purchased or manufactured. Replacement of inventory is essential to continuing in business, and this requires ever more cash. It has been estimated that in a four-year period spanning the mid-1970s, industry overstated true (uninflated) earnings by about $135 billion, of which $85 billion was in understated inventory replacement and $50 billion in understated depreciation. In fact, studies have shown that, in the 1970s, American companies as a whole paid out over $100 billion more in dividends and taxes than their actual earnings.

With the reported profits of American corporations overstated as a result of inflation, the government extracts more income tax from the corporations. This tax, unlike the inflated reported earnings, is very real. To pay it requires the transferring of real cash from the industry to the government. All in all, while the tax brackets have been nominally prescribed as below 50 percent, American corporations in recent years have been paying income taxes of 70 to 80 percent of real earnings. (Of course, the same happens to all individuals. Each dollar of our income during recent inflationary years has fallen steadily in true purchasing power. But our inflated higher income figures have moved us up by IRS rules to higher tax brackets. Thus, with inflation, the government extracts a bigger share of every company's and every person's otherwise discretionary funds.)

Not only the true cash flow but the real return on investment of corporations has suffered greatly in the past highly inflationary decade. From 1966 to 1977 the profits of American corporations as a percent of total national income dropped from over 13 percent to about 9 percent. As compared with 10 to 15 years ago, the actual cash flow available now to corporations for plowback for each dollar of revenue is down to less than half. It is no wonder the stock market is low. Many corporations have a stock market price less than the stated value of their net worth on their accounting books, even though the corporations' assets in plant, facilities, equipment, raw materials, and finished goods could not be duplicated in today's inflated market at any but much higher figures than those book values. Of course, poorer real returns on real replacement value of the investment discourage further investment.

Several other factors are inhibiting risk investments. New technology is by its nature speculative. Risk taking is part of the task of management in the private sector but in periods of diminishing rewards smaller risks will be favored. This means priority will go to sticking to existing technology and making only small changes in techniques of manufacture and in the products themselves. Managers of R&D in American corporations are now reporting a heavy shift in emphasis to short-term programs either to produce safe, non-speculative, incremental improvements or else to learn how to comply with new government environmental and safety regulations. Basic research has been disappearing from private U.S. industry. Another factor is that principal executives, while not disinterested in long-term investments that may enhance the company's position after they have retired, have a natural desire to see results while they are still in the driver's seat. They are increasingly less motivated to make risky, long-term investments as their concern grows about U.S. economic–political stability over the lengthening period required to see a speculative investment through to successful completion.

Even when a company, to play safe in the short term, puts off expenditures to improve the plant and add new products and accepts the negative, long-term consequences of such deferrals, the persistent, rising inflation continually requires additional investment in inventories, receivables, and cash in the checking account. With diminished cash flow, lower real returns on investment, a poor stock market for attracting equity capital, and an inflation that requires steadily more working capital for the same real output, many corporations have overborrowed to raise needed cash. The higher debt to provide the higher working capital is obtainable only with interest costs eating up a higher fraction of the available cash flow. We are seeing facilities in the U.S. go out of date and development held back when it might have created new products and jobs, or decreased costs to fight inflation, or increased exports to improve our balance of payments.

Take the steel industry, so often cited as a field in which we no longer lead. Is it lack of enough management skill, R&D, and technological innovation that is causing our steel industry to drop in world stature? Not anymore, because any valuable process or technique for manufacturing steel used in Japan and Germany could be incorporated here as well. American steel companies are not now choosing to introduce certain advanced technological methods because of anticipated low return on the incremental investment and an inability to raise capital in view of the poor returns.

To all this bad news we must add an encouraging note. No evidence exists that Americans are basically any less innovative now than in previous decades or centuries. "Yankee ingenuity" always has applied to the south as well as the north, to the farmers and city dwellers, to the pioneers who settled the west and to recent immigrants. If there is now less manifestation in America of inventiveness and determination to break new ground, it must be because the pattern and policies of our society today are creating too many obstacles, providing inadequate motivation, setting the wrong environment.

After World War II there was a burgeoning in the United States of new technological industries: computers, agricultural technology, instant copiers, telecommunications, jet transport, semiconductors, nuclear reactors, spacecraft, fast foods, new chemicals and pharmaceuticals, and many more. Some of these product areas are now approaching maturity. We need new ideas and enterprises as well as continued enhancement and expansion of the fields in which we have a strong position. We also need infusion of new technology into such older industries as metals, and whole new developments in energy production. To do this requires a great amount of risk capital and motivation for entrepreneuring and innovation. The private sector must do the job, but the government has to create the investment climate.

United States tax policies more often than not discourage investment in new technology. The government taxes the return on investment twice, once to the corporation on its stated earnings and later again to the investors when they receive dividends on what is left to the corporation after taxes. Out of one dollar of earnings, a corporation may find that 60 cents goes for taxes and 20 cents to the shareholders in dividends (of which the government takes another 10 cents from the recipients). Of the remaining 20 cents, 15 cents must be used to cover added working capital and depreciation and replacement costs above the allowable, reported costs. Finally, then, 5 cents is available for plowback to cut new paths. From future returns that the company gets on this 5 cents of new investment, the government eventually will again take around 80 percent.

We could stimulate more investment in the United States by making the earned income from interest on savings tax-free. As it is now, the meager returns that savers keep after taxes and the high inflation rate discourage them from leaving their money unused. Instead, people figure it is better for them to spend now, before prices go up more. We could gradually decrease the corporate income tax and merely tax the personal incomes of those who own shares in the corporation. Then, if (as in Japan and Germany) we abolish the capital gains tax this will encourage people to invest in new ventures and in general expansion of technologically innovative activities. The government would get its share in increased income taxes from the later, higher returns the investors would receive through dividends. Germany and Japan, each with a zero tax on capital gains, have had the highest rate of investment (expressed as a fraction of GNP). The U.S., with the most severe capital gains tax of all industrialized nations, has, not surprisingly, the lowest rate of investment. The United Kingdom is second to us in severity of tax on capital gains and occupies the place next to our unenviable bottom position on investment.

Our government policies could, but do not now, encourage disinvestment in backward and sick industries. This is as important as new investment in new and competitive activities. If low-productivity operations are closed down, the total productivity will go up as labor and capital are freed to go into the higher-growth, advanced-technology industries. The Japanese have such a policy. They speak of abandoning sunset operations and favoring those in the sunrise condition. The United States, on the other hand, if it does any protecting, is likely to do so for dying industries and reap the disadvantages.

Until a decade ago a large number of small technological corporations were launched in the U.S. each year based on technological innovation. As the 1960s came to a close, initial stock issues to finance such entrepreneurial companies dropped to a trickle. In the past year or two some recovery has occurred, but the founding of new technological enterprise remains well below former levels. Such a falloff should be a source of concern about creativity and competitiveness in the utilization of science and technology in America.

Consider only that the key ideas for the following fields came from individuals who were not employed in large organizations: atomic energy, advanced electrical batteries, computers, cellophane, color photography, cyclotrons, DDT, FM radio, foam rubber, inertial guidance, insulin, lasers, the Polaroid camera, radar, rockets, streptomycin, the vacuum tube, xerography, and the zipper. It is true that the bulk of industrial R&D is done by the top 100 largest technological corporations. However, a loss of the contributions from new, small entities is bound to be penalizing.

The United States government's demotivating policy on taxing capital gains exerts an enormous effect on the starting of new technological corporations. The Revenue Act of 1969, which raised the capital gains tax maximum from 25 percent to almost double that amount (with the aid of additional legislation coming later), had a disastrous effect on new equity issues in the stock market. In 1969 there were over a thousand such issues; they raised many billions of dollars. By the middle 1970s the annual number of issues had dropped by a factor of 100 and the funds raised were proportionately lower. For example, in 1972 there were some 400 underwritings for companies with a net worth of less than $5 million, raising about a billion dollars. In 1975 there were only four such underwritings, raising a mere $16 million.

America does not lack inventors, scientists, and engineers with new technological ideas and with the courage to create a new business venture based on exploiting them. But it has become too hard to get the necessary venture capital because the reward–risk ratio is too low. It is a far more speculative act now than it used to be to start a new company based on innovative ideas. The novel products may be welcomed by the market. However, the cost of doing business will be higher and the return on investment lower than could be realistically assumed in former years. Those who are of a mind to invest in new technology are more inclined now to be leery of the whole national environment. They believe it not conducive to eventual success and are fearful of what the economic situation might be by the time the fresh little company they are considering gets bedded down and its products are ready to market. Venture capital is making a bit of a comeback now, we must note, perhaps in response to Congress' reducing the maximum tax on capital gains from 49 percent to 28 percent. However, unlike the situation of the 1960s, venture capitalists now generally concentrate their investments in companies already in existence and able on their records to show substantial evidence of continued or early success. Leading investment bankers now have a policy of not touching a public issue unless the company has attained a record of earning in the millions of dollars annually for several years.

Surely a great handicap to productivity increases in the United States at the present time — at least when productivity is defined narrowly — is burgeoning government regulation. Whether justified or not, government requirements on industrial operations are now diverting capital expenditures to the meeting of increasingly stricter environmental and safety rules. Severe standards for hazards and pollution control have required large expenditures by industry that do not contribute to earnings. These capital investments also do not increase the output per labor hour. There is no mystery about the fact that when funds are used to meet regulations they are not available to improve the production technology or to invest in further exchange of capital for labor to produce the same output. Also, the production processes must be changed away from the most efficient (by conventional definitions of efficiency) to methods that will satisfy the new regulations. Aside from all the ills of inflation, cash generation for investment in U.S. industry has been diminished by government regulations, the sensible as well as the unnecessary.

Recently, one large chemical manufacturer, analyzing its modernization and expansion program for the decade ahead, figured $3 billion would be required to meet environmental and safety regulations. It also estimated that if the $3 billion were put instead into new plants and products the company has available for the market, this investment could create 20,000 more jobs. More importantly, the company also judged that three-quarters of the $3 billion would be wasted through expenditure on environmental improvement efforts which are misguided, owing to the partial incompetence of the rapidly built-up government bureaucracy and the confusion as to the intent of the laws that set up the regulatory functions.

The Environmental Protection Agency (EPA) spends about $100 million annually to regulate the United States chemical industry and the Food and Drug Administration (FDA) budget is similar to finance the watching of the food and drug industries. Aside from the huge number of chemical compounds used in industrial processes, chemicals are employed to encourage and protect crops, create new food products, and camouflage the wrinkles of old age in food and people's faces. Various chemical products, either old ones or new ones that are coming on the market, are disclosed regularly as being a possible cause of cancer, birth defects, and other ailments. Thus, controls by government are absolutely required. But not all government regulations and the ways in which the government does the regulating are in accordance with the desires of the voting public. If they were we could argue the society gets a good return from the regulatory effort because of the protection it affords. While the realized reward is in a different category from that return seen in the earnings of industry, it is important and often vital to seek it. However, much of the government's activity constitutes overregulation. Moreover, the bureaucratic way in which the regulation is handled makes compliance and cooperation with the government overly costly.

Following highly involved rules, filling in mountains of forms, and negotiating and communicating with the government on safety and environmental controls are requiring large, specialized staffs. Small business often is virtually unable to comply or is hurt badly in costs if it does. For example, compliance with requirements of the Occupational Safety and Health Administration (OSHA) involves much more than the firm's being certain its operations are adequately respectful of the health and safety of its employees and the ultimate users of the product. The government requires a steady flow of detailed information concerning the company's activities. Just to know what is required of it, the company must employ experts in the government regulations as well as in the technical means for meeting them. Also needed are extensive and expensive measuring and testing facilities and, again, specialists to operate them. It is tough to comply with what is right and not easy to satisfy government demands delivered by teams sometimes lacking competence and always requiring much communication and detailed negotiating. Some manufacturing organizations now have to use 25 to 50 percent of their engineers on regulatory matters rather than on innovative product problem solving. Many thousands of engineers in large and small companies are now doomed to a life of handling regulation issues while believing that a substantial fraction of their efforts do not contribute in the end to added health or safety or purity of the air or water.

Improving the environment and increasing safety in the production and use of products are meritorious societal objectives and the government has necessary functions to perform here. However, recent study has indicated that complying with government regulations is costing business and industry from $100 to $200 billion annually. Doubtless some of these expenditures are for safety, environmental, and other purposes totally consistent with the desires of the American citizenry. Equally without doubt, however, is that a program so huge, and built up so quickly must, include much needless regulation. It forces much altering of methods, and buying of apparatus to make these alterations possible, to achieve very questionable or negligible improvements. It also funds huge bureaucracies in both the government regulatory agencies and the industry that must interface with them. We wanted to increase safety and minimize pollution, but we have been slow in working out common-sense balances in the process and have created an effort costly beyond the intended levels.

Without meaning to do so, our regulatory climate discourages innovation in new product areas. Even if nothing but benefits result from compliance with standards, the effect of extensive regulatory controls is bound to be that R&D is concentrated more on smaller risk projects, those where regulatory problems in bringing out the product are less likely to arise. The more innovative a new program, the greater the business risk. More new ideas turn out to be impractical when the details of the real-life market are uncovered than those that deserve to be continued through development to successful maturity. Regulations create additional delays and uncertainties. It is bad enough to have to meet conventional return on investment criteria in an inflationary period, but now the embryonic project being considered must be seen as able also to pass safety and environmental hurdles very tough to anticipate and define well ahead. The easy route is to play safe, put available chips on small improvements, and avoid big steps.

As an example of the way government regulation plays a part in restraining advancing science and technology, consider the field of pharmaceuticals. A very much longer period for testing and approval of a new drug is now required than a decade or two ago. When a pharmaceutical drug company makes an investment in a new product line, it must have confidence in getting through all of the steps, from conception to development to testing and approval and then to start-up, dissemination, and acceptance, before the drug reaches a profitable stage. If the program is stretched out too long it becomes far less inviting to even consider investing in the first step of R&D.

Overregulation in the pharmaceutical drug area may provide added insurance that the public will not be harmed, but it also results in the denial to that same public of the possible benefits. If other countries have more balanced policies, available funds for R&D in drugs and the expertise to develop them will move away from the U.S.

In this chapter we have presented many facts and viewpoints to establish the seriousness of the American technology slip. Much of what is happening clearly need not happen. Surely the trends are in part the consequences of shortcomings and inadvertencies. Some causes of our deteriorating leadership are rooted in government policies that seem just plain wrong. Some government actions have even appeared to a few critics to constitute madness. Yet, as the next chapter will indicate, there may be method in the madness — at least some of it.

4

efficiency isn't everything

We have described in the previous chapter the case that says America's technology slip is real. However, some of this disturbing decline may be characterized properly as the result of deliberate actions seeking and yielding net benefits to the nation. To see why this startling and apparently inconsistent statement may be true let us return to that bothersome trend we first called out, our productivity lag.

All quantitative measurements of productivity in use by economists and others concerned with production involve a ratio of some kind of output to some kind of input. Numerous productivity indices are used, like the amount of grain grown per acre of land, the number of automobile parts produced per worker per year, or the number of tons of output of a chemical obtained per ton of raw material input. These quantitative measures of productivity are all based on the concept of efficiency. Now, the value of any kind of quantitative measure hinges on what use is to be made of the measure once obtained. Productivity statistics are useful if it is efficiency that interests us. It does, but in modern America, to our credit, we are after more than narrowly defined efficiency alone.

When certain productivity measures drop, the decrease may be an indicator of a preferred corollary increase in the quality of life. For example, improving the safety of workers and stopping factories from impairing the environment may result in less goods being produced for each employee-hour expended, lowering both the GNP and the productivity and producing less employee output for each dollar of capital invested in the plant—but yet, this can constitute a "good" for the society. Which connotes a higher level of civilization and human accomplishment: to be able to buy more goods because a productivity increase lowers the costs, or to have the privilege of breathing cleaner air? Which is to be rated as more deserving of a medal of achievement: to produce more goods accompanied by more cancer, or to enjoy less production but better health?

There are other interesting productivity measurement questions. When technological advance makes long-distance calls quicker, more reliable, and cheaper, the society gains in quality of phone service, but the GNP tells only about the sum of all phone bills paid per year. The GNP similarly lists the dollars paid for the automobiles produced but does not credit us for having produced safer, less gas-guzzling, and less polluting vehicles. It is not satisfactory to measure production of goods and services by dollars spent for them, even when inflation is taken into account, and not recognize the improvement in the goods and services being produced.

Often the statistics speak to a standard of consumption rather than a standard of living. This is because the former is technically the quotient obtained by dividing GNP by the number of people—a description of how much buying we do per capita, not how truly satisfying our lives are. To continue this line further, maybe we need to interest ourselves also in quality of working. To speak loosely of a high standard of living could be to ignore that most of us spend about a third of the day at work, a third asleep, and only a third in "living" and miscellaneous activities. Ensuring a high quality of life in bed appears an inappropriate subject for further discussion in this book. On the other hand, it seems fitting to comment that the satisfaction, enjoyment, health, and safety of an employee during the work period perhaps deserves more focus than productivity analysis alone can provide. Inextricably bound up with how productive we are is the question of the quality of what we produce and the safety consequences and environmental effects of how we do the production and use the output—factors related to our social goals and our sense of values.

As to efficiency alone, no doubt exists we are slipping badly. While efficiency is not the only criterion of success, it is important. It is one thing to allow inefficiency because we want to gain something unattainable if we let the goal of efficiency dominate our every policy and action. It is another to be inefficient merely because we are backward or because we have inadvertently put barriers in the way of innovative effort and investment to support it. If we become inefficient enough then we will lose the options to reach the very social goals we chose because we saw them as superior to all-out, blind, efficient technological advance. So we must distinguish between the tolerable or preferred class of inefficiencies, those resulting from deliberate deviation from the most efficient plan, and the useless, harmful, accidental inefficiencies that we should and can do something about.

Trading productivity and efficiency for social and political benefits can hardly be done well if we do not know the values of what we are trading. But do we have good means to effect the tradeoff compromises? On some matters the government agencies that set safety and environmental standards never even consider the related productivity or cost problems, let alone attempt a careful balancing. Sometimes a slight lowering of standards may keep them still within a sensible, safe range while almost eliminating a severe productivity drop that overly severe standards may trigger. If productivity is caused to fall enough, our industry loses out in the domestic and world market, unemployment rises, living standards suffer, the poor get poorer, and malnutrition and illness rise. The resulting negatives to health can be as great as those arising out of carelessness in safety and pollution control. So how do we effect sensible tradeoffs in the United States? While we wish to put off full discussion of this question until later, we might well establish here that it exists. When productivity falls in parallel with actions to add safety and protect the environment this should not be a surprise or a reason for discouragement. It is an example of the tradeoff and decision issues we must expect and learn to handle in a well-run technological society.

Having decided to increase safety for the workers and those who use the products produced, we will have to expend some funds to figure out how to be safer since it is not always obvious. We generally will have to add to the cost and complexity of the facilities and equipment that turn out the product and we probably will have to use more worker-hours to achieve the same output after we make the factory safer. Those funds and efforts could have gone instead into making the operation more efficient, turning out more products with fewer worker-hours and at lower cost. Similarly, if we invest to limit impairment of the environment from production operations, then again we are using resources which could have gone into productivity improvement. Clearly, there is a tradeoff here: greater safety and less pollution versus higher productivity. If the nation's sense of values and balancing of factors has been correct, then we have no reason to be alarmed about the American productivity slip—not as to this specific source of that measure's lag.

A corporation will be interested most often in how productive it is in utilizing its assets and the return on investment will be the end figure it will seek to maximize, not the productivity. One route to accomplish this maximization often is to increase the productivity of the labor that participates in the company endeavor. Given the same physical assets, machines, drawing boards, computers, laboratories, and factory area, if all the employees were to work harder, be more ingenious, make fewer mistakes, learn their tasks faster, be more highly motivated and better suited to the work as to talent and experience, then the productivity (measured, say, by the output per employee per hour) would rise. It could also be made to increase by investing in new tools so each worker could turn out more parts per year or in further automation that eliminates some of the labor entirely. A company that uses 10,000 employees to turn out $100 million worth of products might turn out the same quantity of identical products a few years later with only 5,000 people. This would mean an increase in productivity, the ratio of output to employees' efforts, of 2 to 1 over that period. However, this increase may have been primarily the result of investing many, many millions of dollars in the design and purchase of automation equipment. Without analyzing more details carefully we cannot say whether this big investment is really paying off in terms of lower labor costs versus higher capital, depreciation, and maintenance costs. An increase of productivity by itself is not always as beneficial as the more fundamental optimizing of return on investment.

In earlier cheap-energy periods, industry decreased costs by substituting machinery for labor. Even machinery that was a high energy user generally would increase productivity and improve return on investment. Now the situation has changed. In some industries, notably chemicals, energy has become the highest cost factor. In others, government regulations alone are forcing a change in the tradeoffs among energy, machinery, and labor. When wages increase more rapidly than energy prices and the cost of capital, which was the pattern until 1973, then both energy and capital spending are substituted for labor. Now energy costs are rising faster than labor costs and labor is being substituted for energy. Also, higher energy prices tend to depress the rate of capital formation and thereby reduce the rate of substitution of capital for labor. With this new tendency for rising labor use, the long-term outlook for labor productivity in some manufacturing categories is a slow-down in its rate of growth.

There are many more technology reactions with the society that affect America's technological stature and disclose that productivity alone fails to measure well either progress or deterioration. For example, one way to turn out more and better products per worker-hour is to have superior workers. But what about the untrained and hence the unemployable? How do we change the society so it will not contain a penalizing, substantial, permanent segment of disadvantaged, unskilled, frustrated people? One way is to decide that our productive system must pay the price of including them now as apprentices and trainees, so later they may become full-fledged, skilled employees, and make this a national goal. Surely, this will mean a sacrifice of productivity in the interim, a period stretching out for many years.

Whether individual companies, as a philanthropic social contribution, deliberately hire this less productive labor for training, or the government funds such hiring and training programs using tax money from all of us to subsidize it, the end result on productivity will be the same. In the short term, if we divide the output by the number of worker-hours, this calculation will yield a lower number than we would have achieved in the absence of this social step of equal opportunity employment. However, the program, despite lowering the nation's productivity statistics, should yield net benefits. It must be better for all of us that a worker receive $200 per week in pay, even if the $200 comes from taxing everybody to subsidize that worker's modest contribution during a learning period, than to pay out the same $200 in relief while the worker is unhappily unemployed and learning nothing.

When we alter a manufacturing process so as to use less energy, even if it requires more worker-hours—a substitution of labor for energy—we lose in productivity but we gain in the national goal of being free of dependency on the OPEC nations. Again, we deliberately accept productivity's slippage to achieve other goals. Similarly, we may not want more food produced and marketed if it has to be done by employing particular fertilizers, pesticides, or preservatives whose use would increase the cancer rate. On the other hand, neither do we want people to be deprived of adequate nutrition. That means learning how to balance safety against quantity in food production, a tradeoff for which agricultural productivity cannot be used as the sole guide.

Precisely because hurting productivity may be a necessary, though not highly cherished, corollary to achieving certain high-priority objectives, it is important that we do not impair productivity and constrain technological development where such handicaps buy us no progress whatever toward any perceived goals. If we could develop our creative, analytical, and discriminatory capabilities so as to separate desirable from undesirable technological efforts, we could hope to have our cake and eat it too. The more we can prevent the wrong efforts, the greater will be the human and natural resources available to put behind the right ones. The better we identify and understand benefits and negatives of technological developments, the more successful we can be in working out a combination to meet our goals.

Let us now pause to consolidate what we have so far observed about America's technology status. There is, we have seen, good as well as bad news on the technological innovation front. However, on the whole, the recent performance of the United States justifies our being pessimistic, or even alarmed, about the trends. The precise conclusions to be drawn depend, admittedly, on what ratings we attach to the number of first and second places we or other nations hold and how we interpret the real significance behind any of the statistics cited. To prosper we obviously need not be first in every aspect of science and technology. We never have been. We do need to be outstanding in enough areas to attain our national goals and hold our own in general international competition.

Overall we are still the strongest nation in science and technology, roughly comparable in our total capability for registering successful advances to western Europe and Japan combined. But, however we score the contests, we are lower in standing today against other nations than we were a mere decade or two ago. We have lost leads in important areas and we are in danger of losing more. The other non-communist nations have about the same number of scientists and engineers as we. Those countries are now about equally able to back them up with financial and physical resources from the world at large. Thus, for the future it is not realistic to imagine we can lead in everything or even in most things. The communist countries constitute another large concentration of technical talent. Taking the world as a whole, we should expect to produce only between a third and a half of the future technological breakthroughs, scientific discoveries, and successful implementations of new know-how. As to the Soviet Union, we have noted that they have had the option available to them, and have chosen it, to give such priority to military technology as to be roughly on a par with us, with superiority in some areas and inferiority in others. Generally, they are behind us in civilian technology.

It is also true that the social–economic–political environment of the nation limits our science and technology endeavors, whether it be pure research or implementations. These are severe constraints and they are growing in force. More particularly, we have an array of new national goals regarding safety, health, and the environment that are not clear but nevertheless interfere with productivity rises, take funds out of the innovation-fostering cash drawer and use up capital that would otherwise be available for investment in improved plants and new, more productive methods and machinery. We want to decrease hazards and pollution, advance the disadvantaged among us, and use less OPEC oil. Generally these desires are not compatible with improving our productivity in the short term. The situation does not allow us waste, improper setting of priorities, misguided policies, pressing for unsound goals, or going about attaining them in the wrong way.

If the undesirable trends are not reversed, we will go on losing technological leads and later will be in a difficult, unnecessarily inferior position. Losing technological superiority does not automatically present us with ancillary advantages. We do not need to be behind and there is much to gain by being ahead. You can lose a tennis game by playing badly, but you can also lose it while playing well if your opponent plays even better. In technological advance we are doing some important things poorly—for example, failing to provide the best environment and motivation for innovative effort and allowing adversary relationships instead of cooperation to build up between government and private sectors. In addition, other nations are winning positions of superiority because they are developing human resources, technological expertise, and government–private teaming more effectively than we are. In many instances the United States has opened up the paths on which others are now treading and often running faster than we because they are in better shape.

We have in this country now a valuable new sensitivity to the negatives of implementation of technological potentials. To provide a healthy and socially satisfactory life does not require of us a worship of productivity and efficiency. Neither does minimizing impairment to the environment and health require stopping science and technology advance. On the contrary, effective constraining of the negatives of the technological society requires research and development to increase understanding and lay the basis for further creative effort to diminish those negatives. To preserve options to meet national goals, we cannot afford to handicap our further explorations of science and our advancing of technology through inadvertency, accidental bottlenecks, bureaucratic inefficiencies, and general confusion. Fortunately, as we shall later see, there is much the government, private industry, universities, and the general public can do about undesirable and artificial constraints.

In the United States the principal sources of expertise in innovation reside in private industry and the universities. However, the government is the most potent force exerting leverage on the performance of these primary sources of ideas and useful output. We need to understand more thoroughly the government's involvement in matters of science and technology. Accordingly, we shall examine next just how the government participates in scientific pursuits and technological developments in America.

5

limited free enterprise and pervasive government

The capitalistic, free enterprise system—private funds invested at risk in pursuit of a satisfying return—has served America well in the past to connect the needs and desires of the citizenry with the potentials of science and technology. This wedding has yielded us an immense stream of goods and services. If we are now seriously threatened with a loss of technological adequacy, and if this circumstance is even partially the result of wrong government policies—over-taxing, over-spending, over-regulation, and general weakening of the motivational environment—then why not get the government out of the way? Would not a totally unfettered free market economy fully remove the difficulties of applying science and technology optimally on behalf of the society? The applications then would be chosen by the market, by the criteria and value judgments set by the very customers and citizens of our democracy who make the market.

This proposal is irresistibly attractive to many. Judging from their aggressive commentary, a substantial number of prominent business executives, a few leading politicians and famous economists, a professor or a columnist here and there, and most small business owners are convinced that any problem the nation faces can best be handled by the government's keeping its hands off and leaving everything to the private sector. But in a realistic examination of our options, it is naive to contemplate reconstructing United States society to make possible essentially the removal of government involvement in science and technology. Moreover, in some important, needed areas of technological endeavor now dependent on government financing, adequate investment funds will not be forthcoming from the private sector for the straightforward reason that the anticipated return on the investment, when compared with the risk, is not satisfactory. The financial yield will not be seen as meeting the requirements of the private money market where superior investment alternatives are to be found. As we have already said, inflation and tax policies have worked together to impair real earnings (while causing reported earnings to be overstated) and to decrease the cash flow in industry available for plowback to support technological advance. In the present inflationary environment, the more advanced the technology the more speculative becomes the investment in it and the longer the time required to reach a turnaround (where cash is generated rather than having to be supplied).

But those plugging the concept that free enterprise should be allowed to take command tell us that, precisely because the inflationary environment is caused by wrong government actions, we should stop those actions and government interference generally. However, removing the government from science and technology, even if that would cause technological innovation to boom, would require more than small changes in the overall social, political, and economic pattern of the nation. It would mean altering the relation of the government to the citizen in areas well beyond science and technology matters. For instance, stopping inflation is not a matter of simply "getting the government off our backs," even though many government actions are inflationary. It means firm control by the government of the rise in the money supply, which is politically practical only with major credit, budgetary, and tax policy changes together with a very different set of public expectancies of what government services must be provided. Even with a conservative trend in the country, such a wholesale alteration of government patterns is hardly to be counted on as readily attainable.

Also, an increasing fraction of the potentially most beneficial technological projects calls for investments bigger than the net worth of almost any one corporation, and large corporations in the United States are not allowed to team up to share the risk because of the present practice in interpreting antitrust legislation. For such projects (for example, in energy, ocean resources, space technology, and mass urban transport), the reward–risk ratio is too low, the start-up losses too high, the time to reach an eventual profit phase too long, and the tie between the fate of the program and unpredictable political decisions forbiddingly close. The market for these large-scale projects is far from being free; instead it is a cacophony of semiautonomous and conflicting interests. In addition, many of these programs require a high degree of international cooperation, which is difficult to arrange and requires the government's participation.

Here again, the free market advocates would argue that we simply ought to modify the government's approach—for example, to antitrust—wherever it is the bottleneck. Are such fundamental changes really in the cards when one considers the extent of antibusiness sentiment in the nation today? Many believe the free enterprise sector is not to be trusted with serious interests of the nation. This may be unfortunate because the government is necessary and competent only in certain endeavors and grossly inappropriate in others, and the free enterprise sector has contributions to make which we can ignore only to our detriment; but it is true. Those suspicious of the private sector today cannot easily be persuaded that the government should remove itself from society's most important areas, such as energy alternatives. At the most, they believe that whenever any problem arises the government must step in and solve it. At the least, they see the government as their protection against the enemy, corporations out for obscene profits. Their complaint is that the government is not yet doing enough, that government's involvement and control should be even broader. Even when they join with the growing number of others in the nation who believe the government is an inefficient, overstuffed, incompetent bureaucracy (harboring a few crooks as well), they ask only that the government be improved so these shortcomings will be minimized. They certainly don't want the government out of the act.

Doubtless most Americans are fundamentally anti-government in one sense: they don't want their lives totally run by the government. Also, they may appreciate that free enterprise has something good to offer them. The motivation, flexibility, competitive advantages, and individual freedom of choice which it provides are characteristics of an American way of life they do not want abandoned. Additionally, most of the public would probably tell us that the private sector contains the real core of the nation's expertise for solving certain of our problems. At the same time, however, they insist the government provide an array of services, controls, benefits, and subsidies that automatically limits free enterprise, keeps government spending high, and sets a foundation for the inflationary climate, which with another part of their heads they realize is very penalizing to all of us.

This desire of the people for government control and free enterprise, both at once, is not necessarily bad. Later in the book, we shall argue, it is mainly good. If the concept includes inconsistencies, they are unlikely to be exposed as such by the nation's business leadership. Many executives are quite accustomed to delivering a luncheon address on the benefits of free enterprise and the ills of control by government, then hurrying to meet with government agencies from which they seek contracts, special subsidies, and general favors. Business leaders regularly can be found sounding off about the curse of government regulation while urging the government to regulate their own domestic and foreign competitors so as to give their individual companies an assured sector of the market. President Gerald Ford, on coming into office, started a study to curb excess government regulation of the airlines and immediately found some airline executives at The White House door to express fears that the market structure they were accustomed to might collapse and unfair competition might rise to do them in. In periods of severe financial stress, business managers have been observed more than once asking for rescue by the government.

If to gain more from technological advance the relationship of government to free enterprise in the United States needs to be modified, improvement can occur only by understanding what the problems and the real options are. In a complex, dynamic, multidimensional, technological society that is also a democracy, we have to expect a substantial amount of confusion as to the proper roles of government and free enterprise. Clearly, what we must arrange for is a sensible degree of cooperation between government and the private sector. We always have been a hybrid of a government-controlled nation and a free enterprise, capitalistic one. The trick is to work out the right functions for government and the private sector. The relationship should make sense when tested against such criteria as the maintenance of freedom, a strong economy, competition, flexibility, incentives, protection of the citizenry, fair play, and equal opportunity. Unfortunately, this is easier said than done. Nevertheless, our present disarray is beyond reasonable and tolerable levels. Instead of finding ourselves with occasional areas of lack of adequate cooperation or lateness in defining appropriate roles to make cooperation possible, we have built up almost soley an adversarial, confrontational relationship between government and the private sector on a wide front.

To discover how to make practical modifications in this relationship we can profit by first examining the present pervasive government influence on the science and technology stature and activities of the nation. The government exerts a profound and powerful impact on every aspect of science and technology. It is the chief funder of R&D, the prime setter of priorities, the most demanding regulator, and the leading customer. It sets the rules and creates the environment for accomplishments and failures.

For fiscal 1980 the United States government allocated over $30 billion for R&D, with the Department of Defense using about $14 billion and the Department of Energy and NASA each spending about $5 billion. Of this total, the government has categorized $4.5 billion for basic research: both pure research in the universities and the more advanced research aspects of the mission-oriented agencies just named.

The role of government in scientific research and technological development is multifaceted and controversial on many fronts. Conflicts with the role of the private sector, difficulties with priorities and budgets, problems of having to cater to political constituencies, the incapacitating effects of incompetency and intense bureaucracy — these factors figure into and muddle the government's involvement. As to controversy about the government's role, there is at least one exception. It is now accepted and expected that almost all basic research will be sponsored by the federal government and that the universities are the right places to perform this research. Basic research has become almost totally dependent upon the government because efforts to understand better the underlying laws of humans and beasts, matter, energy, the earth, and the universe are not readily related to specific applications in industry. Even if these most basic of scientific investigations should result in payoffs to the society, it is not through a controllable, preconceived path. Those who set out to perform such research are not seeking patents. They are not after a stream of proprietary products to market so as to yield a financial return on the investment in such efforts. They simply want to understand nature, regardless of whether any consequence of their discoveries could be construed as a benefit. However, the nation has come to regard this kind of research as of eventual value to everyone's health, welfare, and feeling of satisfaction and hence an activity that should be paid for by the society as a whole.

Explorations into the behavior of the universe are not directly connectable with the nation's winning international contests for technological superiority or gaining productivity improvements in the short term. But the results of fundamental research often become foundations for future technological developments. The nation cannot neglect these foundations if it wants continued reinforcement of the well-being of its citizens.

Since the federal government is virtually the sole sponsor of basic research, it controls the patterns and detailed decision making through which the programs are chosen and carried out. In science and engineering, higher education goes well with research that broadens the knowledge base. When university departments carry on research to advance their fields, they also enhance the level of perception and overall competence of the professors and the quality of the students' learning. Through the near monopolization of university research grants, the government thus exerts a strong influence over the education of future generations of engineers and scientists and sets the priorities for attention to specific specialties by the universities. The government is one of the parents, the university the other; together they form the embryos for the birth of future scientists and engineers who will produce further advances in science and technology.

With this de facto role of the government in basic research recognized years ago, one would think the government by now would have developed a smoothly working, objective means for deciding on the size of research budgets and the areas in which funds are spent. Since individual research projects rarely are of one year's duration, it might appear the government would see fit to allocate research funds for several years at a time and would seek to derive for the nation the benefit that a steady and growing basic research program would ensure. Not so. Despite the fact that basic research amounts to only a fraction of a percent of the total federal budget, the total funding level is subject to year-to-year decisions and so is each little grant segment. Doubts about continued backing of their specific research programs create annual crises in the universities.

Admittedly, deciding on exactly the right magnitude of pure research funds for the United States would be a formidable task even if no political issues entered the deliberations. Numerous imponderables always must be figured in, and experts do not always agree as to which fields are most deserving of advance. Differing value judgments by scientists and others lead to varied ratings of the potentials of competitive research projects to advance differently perceived national goals. Political factors are present, such as the pressure from some members of Congress to deliver government funds to their area colleges regardless of whether their staffs have competent researchers.

Granted these difficulties, we would expect, nevertheless, that because of the recognized past benefits to the society from pursuing basic research, the United States would find it sensible to err in the direction of mounting a generous program. A sound rule, for example, might be that university researchers with proven competence as judged by peers would find it fairly easy to obtain reasonable sponsorship. At the least, an outstanding, internationally acclaimed professor, we would think, would be assured of having the funds to provide for laboratory expenses, a technician, a post-doctorate associate, and a few Ph.D. candidates working on their dissertations. We are not speaking here of a university project headed by an overly eager, inexperienced, empire-building professor who wants to arrange something like a manned landing on Saturn at a cost of $50 billion. Instead, we are referring to annual grants in the range of $100 thousand to $200 thousand each to support perhaps a thousand small research teams, for a total of around $200 million, a hundredth of one percent of the GNP.

Unfortunately, our government does not operate in this fantasized pattern. The present funding level leaves good research talent not fully used. Moreover, the government's practices force too large a fraction of the time and energy of those researchers who do get funded to be occupied with selling and administrative chores that did not exist earlier. Federal funding for basic research in the universities today is not adequately covering even the existing tenured faculty. With bureaucratic administrative procedures and accounting requirements using up funds and inflation increasing the cost of the equipment and supporting services, the basic research establishment of the nation, essentially the universities' research activities, is becoming weaker. The universities' inability to utilize research talent fully is especially penalizing to the nation because their budgets do not allow for growth of that talent. More than half of the promising young people earning doctorates in science are finding it difficult to obtain faculty positions that allow them to parallel teaching with research. We stand in danger of aborting the research contributions from the rising generation of scientists. Particularly needed is more support for predoctoral and postdoctoral scholars. The shortsightedness of our one-year-at-a-time, stingy budgeting policies is not duplicated by any other industrial nation in the world.

Admittedly, the government must be conscientious in seeing that its grants to universities are employed in accordance with the government's rules, with no fraudulent intent or careless waste. For this purpose, it must insist upon adequate documentation. Proper supervision of the government's funding must start with a clear understanding of exactly what the government grant is for. A decade and more ago this beginning agreement used to be arrived at in a quick, common-sense fashion. The paperwork has now escalated to where it is an enormous burden on the university. In the interests of being sure that government money is not stolen or misused on foolish projects, there is now a good deal of pointless dissipation in costly preliminaries before a grant can be obtained.

Scientists know that what will be discovered in pure research explorations cannot be promised or described ahead of time. Yet government procedures now have the effect of pressing the research-ers to sell themselves and their projects by pinpointing the relevance of their investigations to specific, readily justifiable goals. For instance, it is easier to get an appropriation for any biology-related project if it is hinted that the work may lead eventually to a cure for cancer. The procedure now requires more presentations and negotiations be-tween researchers and the government. The researchers must explain what they hope to accomplish and how they will go about it to people in the government who often cannot understand why the research is being attempted in the first place. Reviews by peers, through volunteer committees or the like, used to be sufficient for the selection process. Now they are not. More than one senator or repre-sentative seeks to make a reputation as a guardian of the govern-ment's purse by announcing the discovery of funding of a research project that sounds silly. Prominent legislators have been known to decide which projects or scientists to belittle by reading only the title of the project. That title, a five- or ten-word description of what a researcher is seeking to accomplish, must now be designed to read convincingly—that is, imply readily understandable, practical poten-tial benefits—to a skeptic and often to an influential legislator intent on taking advantage of the fact that research will often be judged a waste by those ignorant of the scientific approach.

Government regulations cover many fields now and universities must deal with these requirements before signing research contracts. To do so, universities have had to add administrative staff, including lawyers, accountants, negotiators, and personnel experts. These pro-fessionals in the laws of equality in employment, safety hazards, government auditing, and other regulations cannot apply their exper-tise to an individual grant as it reaches contract stage without substan-tial joint effort with the research scientists on the program being negotiated. When it is said that the government is increasing its appropriations for research in the universities, it is commonly over-looked that a growing fraction of the total funds is going into these non-research activities. They use up the time of the researchers, who have fewer hours to pursue science.

Aside from pure research in the universities, the government dominates important areas of applied science even when the funding and decision making are partially in private hands. In focused, mission-oriented R&D seeking specific short-term results, the government's budget for defense, energy, biomedicine, space, agriculture, environment, transportation, and other areas is roughly the equal of all privately financed R&D. Most of the government-sponsored effort is performed by private entities under contract. Through such contracting the government is the largest force in advanced industrial technological development just as it is in basic university research.

Take military R&D as an example. The jet airplanes operated by the commercial airlines have come from aircraft companies whose engineering know-how in designing and producing such carriers has been based heavily on their work on military airplane contracts. Government-financed projects have led to detailed advances in aerodynamics, structures, controls, communications, propulsion, and other techniques applicable to both military and commercial planes. Military projects have helped provide factory volume and thus shared the fixed charges of producing commercial planes. Without such assists, the civilian planes would have cost too much to justify their use by the airlines or would have arrived at much later dates. The aircraft industry, which contributes greatly to United States exports and which, on the whole, leads the rest of the world, continues to depend on the government as a vital source of support. Moreover, as was proven a few years ago when the government guaranteed bank loans to rescue Lockheed, the government is a financial backer of last resort to a large aircraft company.

Airplane technology, of course, is not the only area which military programs have helped to expand. Guided missiles developments have advanced the technology of automated control of all manner of electrical and mechanical equipment. The solving of military technology problems in materials, combustion, heat transfer, precision measurement, and many others has been important to all technological advance. Military technology has greatly expanded the computer and communications engineering know-how of the nation. In space technology, the military launches the most complex satellites (for reliable and secure, coded, multichannel communications and for intelligence and reconnaissance). Military requirements typically tend to be more severe than civilian needs and push the art into regions where civilian applications become economical and practical only later.

The strong impact of military efforts in all technological specialties can be expected to continue into the future. Even if we should find ourselves suddenly and surprisingly entering a world era of broad multilateral disarmament and greatly reduced expenditures for the military, a need would exist for a continued extension of technology to make possible worldwide arms inspection on land and oceans and from the skies. Complex information networks would be required using the latest in computers and communications technology so as to provide the world with knowledge of any activities that might constitute violations of the disarmament rules.

Military R&D has for many years occupied about half of the best engineers and applied scientists in the country. Aside from the interactions existing between advances in military and civilian technologies, the military effort utilizes brainpower that might otherwise be available for commercial developments. The majority of technological corporations in the United States carry out some R&D and production for the military along with their commercial activities. The government's control over military technology thus is a major dimension of the government's influence on all science and technology. All in all, then, the continuous search for superiority in military weapons systems has pushed forward almost all scientific and engineering frontiers and focused and influenced the use of our available technological resources, both human and physical.

There can be no free market, in the real sense of these words, in weapons systems technology. No private corporation, based upon its judgment of the future military market, is going to develop with its own funds a new advanced intercontinental ballistic missile or a system to shoot down enemy missiles or even a new tank with the idea that at some later time the government will see the need for and decide to buy the developed equipment. The market is made by dangers to the national security and the government's interpretation of these dangers. The strategy to deal with them is government-formed and the process is highly political. Military requirements stemming from government strategy and policy and the technology to satisfy these requirements must be brought along together by government initiative. The risk factor in large-scale private investment seeking to create a proprietary position in a future military market will permanently be regarded as too high by the leadership of private industry. It follows then that military technology is equally permanently a government-dominated, not a free enterprise field. The role of the private sector in taking on responsibilities for R&D and production through government contracts offers ample opportunity for risk taking and for making those contributions most suited to industry.

One key aspect of military technology is the development of nuclear bombs. No product could be less a candidate for the free market arena than nuclear weapons. That the government must continue to be in control appears to be beyond discussion. Nuclear weapons technology advance has an enormous influence on all nuclear technology, including especially nuclear energy, whether through conventional light water or breeder fission reactors or the more exotic thermonuclear fusion processes. Because of continued military technology requirements alone, the government will continue to be a powerful force in nuclear energy.

There are, of course, other reasons for the government's involvement in energy. Already the cost of the government's programs in the science and technology of energy, including nuclear energy, synthetic fuels from coal and shale, solar energy, and others, exceeds the total of all of the advanced R&D now privately funded in the field. Since later chapters examine the government–private industry relationship in energy in some detail, we shall only say here that what has happened in the nuclear energy field to date has been under almost constant direction by the government. As to other forms of energy we shall see later why it is prudent to assume that because of what the government will choose to do, the money it will spend to do it, and the general social and political environment, the government will continue as the dominant factor in energy, be it synthetic fuels, solar, biomass, winds, tides, hydroelectric or whatever.

A good example of the nature and breadth of government involvement in energy is shown by the proposed Alaskan pipeline to carry natural gas from Alaska, through Canada, to the lower states. The cost is estimated at over $20 billion, and the frigid region in which much of the pipeline will have to be built implies technological innovations and adds a great deal of speculation as to the eventual cost of the project and the date for completion. Attempts have been going on for years to put together the required capital from private sources. However, the project cannot proceed without approval from numerous American regulatory agencies, and progress on the project has been stymied by the needed negotiations. Also, the government has the decision power on how much the gas can be sold for and, thus, how much investors in the pipeline can conceivably earn on their investment. In addition, the government is the boss on environmental standards, which increase in severity with time and require more science and technology advances with each resetting. Under these circumstances it is easy to see why the project may not attract the investment funds required from the private sector. If this pipeline is eventually seen as urgently needed, it may have to be built as a government-financed and directed program.

Beyond energy, the government's participation in matters of science and technology in the United States is now so vast that one has only to list the specialties of science and technology and the government's present and almost certain future participation comes immediately to mind. Automobiles? The government, through safety, air pollution, and gasoline MPG (miles per gallon) regulations, has become a partner in the design of the car. Transportation—urban, railroad, or airline? Government funding and government regulations are decisive forces in setting the rates and investment bases for science and technology advances applying to these fields and the priority of the directions of such efforts. Space technology? In addition to military applications, the government funds research spacecraft to investigate the earth's surroundings, the emanations from the sun, and the characteristics of the other planets. The government created and controls the space boost vehicles and facilities used to launch all spacecraft into orbit for military, commercial, and research purposes. The first corporation engaged in satellite-based communications services, still a leader in the field, was created by the government. Medical and environmental research? The government is the largest operator of hospitals and provides the major funds for medical care and for research in the causes of disease and their cures. Through its control of drugs, the government greatly influences R&D in all pharmaceuticals. The oceans? Science and technology related to the oceans is almost totally a government-funded activity, as is weather prediction and control.

In many important technological areas the government has the role of deciding who is allowed to participate. Through licensing of privilege and setting of rates and standards, the government exerts control on the entire communications industry, including telephony, radio, broadcast TV, and cable TV. The government runs the patent system which grants to inventors exclusive proprietary rights to specific technological ideas. Through antitrust and other laws setting up antimonopoly and fair competition policies, it directs various companies to stay out of some areas. It controls the rates and hence the investment and expansion policies of utilities that provide electric power and transportation services. Through the Postal Service, the government affects developments in the field of information technology because competitive ways to move information exist and how the Postal Service meets information-handling requirements and expects to use advanced technology to do so in the future is a key factor in the private sector's progress in information services.

The government, we have noted, grants patents which give inventors special privileges. If it had been written a few decades ago, this book would have been justified to say that private investing to create technologically innovative products needed to go hand-in-hand with a conscientious and zealous policy of filing patent applications on nearly every conceivably new aspect of the science and engineering work covered. In that earlier period, experience and promise suggested that with a patent an inventor could hope to dominate a field of endeavor, keeping others out or extracting royalties from any users of the patent, and generally enjoy an advantageous position. Patents constituted an important motivational influence on technological innovation in the past.

Today one would have to qualify such a summary. The value of a patent is much more doubtful now — an example of how present inadequacy of motivation and diminished hopes of good returns on speculative investments handicap the innovation process in the United States. The complexity of government patent management has increased the cost of obtaining a worthwhile patent. But what has really hurt the value most to the inventor is the government's mood. In ensuring that no improper monopolies will exist, the government has found it difficult to view a strong patent position with favor. Of course, a patent may not be sound in the first place. Any innovation usually is related to or an extension of existing art. A detailed, competent study to define precisely what is new about an idea and distinguish it from numerous earlier or similar inventions takes understanding of the technology and research on the specifics of both the individual proposal and the competitive ideas. The government cannot readily provide this kind of analytical effort and it does not furnish a guarantee when it grants a patent. The patent applicant may not have been first and the concept may not be new. Patent ownership sometimes can be decided only after long and expensive court actions or the settling by the government of interfering claims as it works toward a decision to grant a patent. Resolution by the government of an interference between conflicting claims of two would-be patentees may go on for ten or more years.

The confusion existing today between the basic concept of monopoly, which a patent grants an owner, and the anti-monopoly laws of the nation is especially unfortunate. Often a corporation that owns a patent is well advised to allow others to manufacture under the patent with only a modest royalty payment in return for the privilege, in order to avoid difficulties with the government on alleged violations of antitrust laws. What this all adds up to is that putting money behind a patentable idea to exploit it is less attractive than it used to be. The idea now has to appear to be a real humdinger. The standards for common-sense investment have become more severe.

Small business, so vital to overall national technological innovation, is especially hard-hit by the present patent system practices. Not only has protection for the holder of a patent eroded; with the examination, judging of interferences, and final issue of a patent having become more complicated, the high cost and long time to finally get such coverage as does emerge make the patent less valuable to the small company that seeks it. Even worse, the company has to be prepared for the possibility of long, expensive litigation to defend the patent if the competition is strong and possessed of similar patents.

Since a large fraction of the R&D in our country is government-sponsored, it is important to ask about the motivations for innovation in such efforts. With most of the government R&D programs carried out by contract in universities or private industry, who owns the patents that are generated? Who is permitted to gain advantage from participation and thus encouraged to participate? The government's huge R&D budget presumably is being spent to learn something important and to make use of what is learned. Generally, the government holds the patents instead of automatically granting the full rights thereto to the inventors, the assumption being that to allow exclusive rights to go to the inventors would be to enrich them at the public's expense.

One way to judge whether we are getting a good return on this government-financed effort is to look at the evidence of value in the patents generated, since the government operates on the proposition that the patents represent real worth. Practically none of the 5,000 to 10,000 inventions arising from government-funded R&D each year ever sees the light of day in commercial application. There are now over 30,000 government-owned patents that nobody seems to be interested in taking up for commercialization. Only a few percent have been licensed and only a fraction of these ever made it to market. The costs of producing and marketing an invention are much greater than the original costs of the R&D leading to it. Accordingly, not many private risk takers are interested in putting up the funds required to fully launch an invention unless they have a good chance of exclusive rights to it. For government-sponsored patents to result in more benefits, some scheme to motivate private investment is needed.

We called attention earlier to the fact that a growing fraction of all potentially beneficial advancing technology requires large capital risks and long time periods before the program matures and provides a return on the invested capital. Creating a new commercial air transport or synthetic fuel alternative to petroleum or urban mass transit system or major pipeline can easily require several billions of dollars, a magnitude exceeding the net worth of all but the few largest of American corporations. At the same time, such gigantic projects inevitably involve the government one way or another because large programs are bound to have large social, economic, and political repercussions. The unpredictability and anticipated capriciousness of governmental actions, the high risks associated with the magnitudes of required capital, and the long turnaround times combine to ensure that the really large projects won't be launched by free enterprise acting alone.

Quite apart from a reluctance to mount mammoth projects, large corporations with a dominant position in one specific area of technology have to be very careful about pushing their position of leadership too hard. If a company already has a half or more of the total market in one field, with each of its several competitors smaller, that market leader has a tremendous business advantage. Its higher volume usually translates to lower unit costs. The fortunate company also can cover basic expenses and have more left over to plow into R&D to improve both its manufacturing techniques and its products' performance. Because the dominant supplier should be in a position to outstrip its smaller competitors with technological advance, this could cause the leader to gradually trend toward a monopoly position as some competitors drop out. The company that enjoys this business success probably will be viewed by the government as in violation of the nation's antitrust laws and a candidate for a forced breakup. The threat of this government action causes the lead corporation to limit its degree of advance in its prime area, perhaps doing some of its investing away from the sensitive field. The nation thus gets less than the best effort to advance that specific area of technology. When government action penalizes companies that reach positions of competitive advantage as a result of successful innovation, it obviously discourages innovation.

On the whole, other countries do not have these rules. They deliberately and regularly use many teaming combinations — companies with each other, or favored companies with the government — to achieve technological advance. In Europe and Japan, governments actually take steps on occasion to press certain of their companies to combine for added strength against the United States and other foreign entities. Our policies, in other words, while based on the meritorious concepts of increasing competition and banning monopolistic practices, have the effect of limiting risk taking and weakening our private technology sector vis-à-vis foreign competitive entities.

The government of the United States does deliberately approve certain classes of monopolies. For instance, it allows telephone and electric power companies to be exclusive operators in assigned districts and it licenses television and radio station operators to have sole access to selected parts of the electromagnetic wave spectrum in certain geographical regions. While seeking to ensure a degree of competition, the government also assigns to airline and railroad transportation units certain special positions not available to competitors along particular regional corridors. Unfortunately, however, although not surprisingly, government regulation of communications, power, and transportation utilities focuses on rate controls. To ensure that high profits are not realized, the process often arranges that losses are common and these poor financial performances satisfy critics hypersensitive to profiteering. But the process also guarantees little private investment in new technology to improve the systems. When combined with wrong policies as to depreciation in an inflationary economy, the allowed rates are often too low. They do not permit the utilities to build up capital for maintaining or expanding the operation, let alone providing plowback for R&D to improve it.

Many decades ago, American railroads represented an exciting arena for new technological developments. This is still true in other leading technological countries. Japan and Germany, for example, are pouring money into advanced technology, seeking practical train speeds of 200 to 300 miles per hour. Both countries are optimistic about the magnetic levitation approach in which the train is lifted by magnetic force so that it flies in the air just above the track structure. The United States has only tinkered with railroad technology and is far behind. However unintended, over-regulation has brought our technological innovation down to an essentially zero level for the railroads. The possibility exists that this will also happen to the airlines. They can be held back on rates by the government and caused to suffer a poor return on investment even as the growth of the movement of people and freight by air is high. The airlines will then be in no position to make commitments for superior, more economical, less fuel-dissipative, environmentally purer, and safer aircraft. The technological corporations that invent, design, and produce the carriers will find it difficult to acquire the funds to take on the risk of developing aircraft for the future. In time the government will have to increase its influence on commercial air travel because it is a basic necessity for national economic and security purposes. Again, the overall result is demotivation with regard to technological innovation.

A further example is telephony. It was recognized early that it would be chaotic and ridiculously expensive to allow more than one telephone company to share a geographical area, competing to put telephone receivers tied to different central switching systems in all homes and businesses. If the country had not adopted the idea of a government-granted monopoly to provide telephone service, we would all have needed to proliferate the installations of equipments at our businesses and homes in order to be able to reach other businesses, friends, a physician, or a stockbroker, and the service would have been much more expensive because of the duplication of the switching stations and all the interconnections. The monopoly approach is sensible, but with it comes the reasonable requirement that the government control rates to be paid for the service and periodically examine the return on investment to see that two things happen at once: the telephone company has a reasonable return so that it can raise the capital to maintain, expand, and improve the system; the public is not taken advantage of and is charged a proper price for the service it gets. This kind of government control is high in leverage. Unfortunately, with over-zealousness in holding down rates, investment in improvements can be halted.

The predominant company in this business is the nation's largest industrial corporation, AT&T. With its Bell Laboratories for R&D and its Western Electric subsidiary to produce equipment, it has led the world in providing beneficial technological developments in telephony. But the government is now carrying on an antitrust suit against AT&T, seeking to break it up by separating from it the Bell Laboratories and Western Electric. Granted the best of intentions on the part of those in the government who made the decision to sue, the endeavor nevertheless is the opposite of what is required for technological leadership in telephony. If AT&T's R&D and equipment manufacture both are separated from the operation of its telephone system, its overall effort will drop in quality. It will be harder to continue to provide for the right selections for technological advancements in telephony and the progress will be later and more expensive to attain as compared with what would otherwise have been available to the nation. The government has ample means for protecting the society against the evils of monopoly merely by requiring that AT&T, in its procurement policies, adequately consider sources of equipment from outside as well as inside its organization and license its advanced technology to smaller corporations at reasonable fees.

If through avid pursuit of its interpretations of antitrust the government succeeds in protecting us totally against evil monopolies, it may at the same time cause us to lose technological superiority. This will deprive the United States citizenry of the benefits of technological developments which would otherwise take place. We should not have to choose between accepting all of the bad, or, in precluding that extreme, failing to derive any of the good. Unfortunately, the government is not organized for balanced tradeoff decisions.

From these specific influences of government let us turn now to the general effects of government on the environment for technological innovation. As we have already cited, developments in technology are affected by government policy on the economic front. An example is inflation, which has the greatest impact on technological innovation and concerning which the government has heavy responsibility. Another broad area of interaction between the government and the private sector increasingly has become a serious constraint to the best employment of science and technology.

The United States government is a large and complex structure. What we are about to list as a characteristic of it is not true of all its departments, and certainly not of all individuals in them. However, in the legislative and executive branches of the government a strong prejudice against the private sector is common. Part of this bias, not surprisingly, is merely a form of rivalry. At the level of the individual, the private and the government sectors constitute two conspicuously different career paths and they are constantly being compared without objectivity. The Civil Service system is frequently described by those in the private category through such words as incompetence, inefficiency, laziness, empire building, and bureaucracy—verbiage that is scarcely intended as complimentary. It is common in the private sector to suggest that individuals in government, from minor office holders to top executives, could not earn the same salaries in competitive industry because there the requirements are more severe and the competence level is higher. The job-for-life situation of Civil Service often is resented by private industry workers. Strikes, taken for granted as normal in private industry, are regarded as unethical and traitorous if indulged in by government workers. Industry people, in turn, are often regarded by government employees as the enemy, ready and eager to try to get away with whatever possible, looking for loopholes around the laws and given to evasions as well as avoidances.

To add to the segmentation that is the basis for some of the mutual prejudices, large segments of the government are involved in the regulating of business, directly or indirectly. The Congress passes laws creating agencies to set standards and report deviations from them, so government employees become police. This creates a mind-set of "the good guys and the bad guys" on the part of the government workers and industry employees as they interface with each other. Industry employees who are closest to government, supplying information, and negotiating and defending against accusations and challenges, look upon the government employee as the accusing provocateur. They are insulted by the government's suspicions and believe its employees go overboard in generating hardships to handicap business.

People being people, there is empire building everywhere. In the government some individuals will inevitably be dominated by pursuit of power. Often they will envisage any loss of government control over details of the nation's operations as a threat to their own careers. As they see it, improper and steady private sector pressure seeks to limit the government's influence and a constant contest is waged between government control and private independence. For individual government employees at low levels, these perceptions influence their careers, self-respect, and satisfactions. Many elected officials at the top of the government structure also see their political careers as dependent on the outcomes of private–government control battles.

Now that we have mentioned the top structure of the government as a participant in the rivalry of government versus private control, we should not overlook the role of the President of the United States. From Kennedy's confrontation with the steel industry to Carter's with the big oil companies, we can observe an anti-business and anti-profit theme continually entering the existing presidential rhetoric. Why is it that presidential statements never cite low returns on investment but often declaim isolated high-profit figures which are misleading out of context? A big company with huge investment can have high reported earnings and yet be in financial difficulty. Apparently it is politically unattractive for a president to recognize this. The failure to use the office of President to put before the people facts that have to do with profits and return on investment in the free enterprise system is the same as propagandizing against the system and further evidence of an anti-business bias in government, here at the top level.

The important drag on innovative effort in the private sector is certainly not a lack of creative, inventive talent. The problem is that the government sets the environment for technological innovation while, at the same time, innovative capability in the United States resides almost entirely in private industry. To substitute a harmonious partnership for today's contesting is not easy, and not totally required, because no single formula for definition of best roles and missions will suffice for all areas of science and technology. But we can at least cite some simple rules for guidance.

For example, we should endeavor to keep the government out of those activities where the free enterprise system can do a better job. The government has to be a permanent factor in setting the climate for technological innovation but has little to offer in directing technological and scientific effort. Most of the experts are found in private industry and the universities; relatively few are in government. Now that it is realized that innovation may bring not only benefits but penalties, we must create the interest and motivation to eliminate or minimize the bad while accentuating the good. This cannot be done by government regulation alone, by merely defining and prohibiting that which is bad. Very often we can arrange to have most of the potential good with negligible harm, but that usually takes more understanding, more R&D, more innovation, and more balancing of trade-offs. If the government is to control dis-benefits from technological activities, it must possess, and continually advance, knowledge of these harms. It must sponsor research to establish the base for this knowledge.

The government must be involved in many areas of R&D where free enterprise won't or can't do the job alone, including the sponsorship of pure research in the universities. As to university research, the government should set up more generous budgets, because there is much more to gain than to lose by such generosity. The funding should be committed long-term and there should be a minimum of administrative burdens or trys for zero waste in carrying out the research. The government policy should be to minimize the number of talented, competent researchers in universities who have inadequate support.

Beyond pure research support, it is necessary to be very careful in stating the preferred role for government in R&D. In seeking the soundest applications of science and technology, we should recognize that R&D is only the tip of the iceberg. Other factors, such as matching the technology to the market and arranging efficient production and distribution, are often more important than the magnitude of R&D funding. We need expert handling of these other aspects — something the private sector, not the government, is best able to do.

Thus, if we see a growing weakness in the United States in some area of technology, we should not jump to the conclusion that a new government R&D program is the answer. To be successful with technological developments, the ultimate customer has to be reached and must be satisfied. That customer, for most manufactured goods (not missiles or nuclear weapons), is well represented by the market—not by a government contracting officer or a government technology development manager. The market will make the best decision as to what R&D should be accomplished and what products represent success in that area of technology. The government, for such technology, is at best an incompetent third party standing between the private sector and the market.

The technological innovation slip in the United States is seen by many as an area deserving high priority for government action and equal in importance, say, to the energy situation. Some, believing the government must provide answers to all problems, will dream up expensive government programs and new bureaucratic agencies in an effort to reverse our slipping technological status. The costs and interference with the free market from such activities could further impair our technological strength rather than enhance it.

As an example, it has been urged that the government provide special tax immunities and special tax credits to share the cost of R&D in industry whenever such R&D leads to a "breakthrough." Let us analyze this suggestion. Take a typical technological corporation employing a thousand scientists and engineers inventing and improving products and seeking ways to lower the cost of manufacturing them. This company now notes that the government has announced a tax advantage will be granted if the company can convince a new government agency that breakthroughs have occurred. Most technical effort of corporations constitutes minor improvements and evolutionary changes. Such R&D is clearly not to be eligible for the financial rewards the government would offer in this proposed scheme. In a few instances, real breakthroughs will be made and these will be so obviously deserving of that title that little difficulty will be found in winning government recognition of them. But most candidates presented for government approval—considering that it will be in the interests of the industrial corporations to be liberal in offering them up—will lie in a gray area between the obviously non-breakthrough and truly breakthrough items. Proposals will be written by industry on large numbers of such negotiable items. Each petition will have to be studied by and argued before some kind of a government judging group endeavoring to deal with all the specialties that make up the nationwide technology spectrum.

Imagine the huge number of engineers this could involve within industry. The added load on the industry payroll would either increase prices to the consumers or decrease the return on investment and thus discourage further investment in technological innovation. But the most absurd aspect would be the government's new staff requirements. To do its job, the government would need experts in all the technological disciplines, engineers who can sit down and negotiate on some reasonable level of competence with the experts in industry to judge whether the project being discussed falls into the breakthrough category or not. The general level of government incompetence to accomplish such a task would have to result in arbitrary and confused decision making. The expenses would be borne by the taxpayer and there would be that much less to spend on real technological advance.

Consider another misguided proposal for government action to aid technological innovation. With an enormous worldwide automotive industry in existence employing 100,000 or more experienced engineers and scientists, a congressman recently introduced a bill, "Automotive Innovation and Productivity Act of 1979," to establish a five-year federal program to develop a new automobile. The bill would tax each car sold in order to fund a new national automotive agency to spend the funds collected and attempt to do a better job of designing a car than the competitive automotive industry. Already, 20 manufacturers in America, Western Europe, and Japan are feverishly competing for the auto market in the United States. What, beside costly amateurism, can the government contribute in design, manufacturing, or marketing expertise?

As an interesting further example of the wrong approach in government–private relations to spur innovation, we should note the announcement in January 1980 that the Energy Department and General Motors had signed an agreement to share $65 million to make and demonstrate a gas-turbine version of a Pontiac Phoenix by 1985. If GM had thought the approach a really good idea it obviously could have elected to go ahead, without need for the government financing. The government can contribute only money to the project, but the money comes from taxing GM and other companies and the purchasers of cars, and is diluted by the government's administrative costs in making the transfer.

Whenever the government creates a new technology program with government funds the money has somehow to come out of the private sector. Even if it is a thoroughly justified project, the assignment of funds to it decreases the available investment for the rest of the economy. It injects the government to give one group of private borrowers or R&D teams a preferred position in the marketplace for investment funds over other private entrepreneurs. Having done this, the effective interest rate paid by all other borrowers and users of investment funds has to go up.

If the government is often not suited to direct technological innovation efforts, there are also roles equally unsuitable for the private sector. An example is setting standards for safety, health, and environmental protection. Free enterprise certainly has a part to play here and the United States has not handled well the furtherance of high motivation for the private sector's role in this area. But no matter how capably we shape the free enterprise participation in safety and environmental controls, a mission remains that the government must retain. We should expect confrontations on safety, health, and environment between private industry and government, the former focusing heavily on trying to realize a good return on investment, and the latter seeking to cause private industry to adhere to standards set by law.

It is not that free enterprise wants to violate the law. Rather, it is not industry's function to judge the degree of gamble the nation should take in the tradeoff of costs versus health and environmental controls. Here, presumably, the citizens as a whole should do the deciding, making their wishes known through the election of their representatives, who create the rules for regulatory activities. Advocates for further government involvement with science and technology matters argue correctly that there is often a big difference between what private enterprise will or won't do in technological developments and what is socially desirable. Industry is not a substitute for the government in most regulatory matters. But the government should avoid regulating where free market forces could best provide the answers. Whenever the government controls prices, for example, or allocates output and creates subsidies, it is interfering with the useful workings of the free market. This should be done only when the free market will not suffice to meet national goals or when dominant non-market factors are inevitable and are already keeping the market from being free.

C. Jackson Grayson, Jr., who from 1971 to 1973 was chairman of the government's Price Commission, is not alone in what he has been saying. Having studied 40 centuries of wage, price, and allocation controls, the nation's more recent involvements with energy and wage–price guidelines, and his own personal experiences in directing controls for the government, he recommends: remove all allocation controls over gasoline and other fuels; remove all wage and price controls; abolish the Council on Wage and Price Stability; return to a market economy. He cites the classic example of West Germany. In

1948, that nation removed virtually all economic controls including price and wage controls specifically. Fears expressed in many places of what this would do to Germany (greater inflation, more unemployment, political instability, enlarged problems of the poor, etc.) proved unsound. Employment rose. So did real wages and productivity. Black markets disappeared. After an initial spurt in prices, inflation was reduced, production became abundant, and productivity rose rapidly, a situation that has existed in Germany now for a third of a century.

We have seen in this chapter how pervasive and powerful is the role of government in science and technology. The government funds, sponsors, sets rules, regulates, assigns privileges, directs programs, and restricts entry into numerous technical areas. At least as important, the government creates the environment in which technological innovation occurs and in which private investment takes place. For the United States, a free enterprise market acting alone, with the government a mere bystander, is as politically unreal a system for producing a match of technological potential to the public's needs and perceived wishes as is the other extreme, a totalitarian government, with the private sector abolished. For the foreseeable future, the workable organization for America is a hybrid of a free enterprise and a government-controlled society. The plan must be to assign the right roles and missions to the private sector and the government.

No simple rule exists for doing this. The dividing of responsibilities and the scheme for cooperation must be accomplished to suit various situations and areas. Energy, communications, environmental protection, urban transportation, agriculture—for the different fields the proper division of tasks will be different. Even within one area, say energy, the allocation of roles may vary. What the government and private sources should best do in nuclear energy is not the same as their most effective respective assignments in syn-fuels, for example.

Thus the central portion of this book, which we have now reached, will deal with a number of specific critical areas of technological advance, the aim being to uncover the optimum government–private relationship for each. Before we do that we must digress for a chapter to call out a missing function needed in applying technology for the society's benefits while at the same time minimizing the negatives. It is a vital function, one the free market will not and should not, and the government now does not, provide. We need to explore this vacancy so that the following chapters can be based on an adequate understanding of existing organizational bottlenecks.

6

the missing decision department

What is the most severe constraint on the advantageous, timely, and prudent utilization of advanced technology? Is it inflation, because it so weakens the economic environment required for investment in innovation? Inflation distorts and impairs every beneficial human endeavor, but its harms are neither special to technology nor focused on it. A more powerful deterrent to exploitation of technology is peculiar to technological applications: disbenefits are produced along with benefits.

Set up any process, build any machine that the laws of nature permit, and no matter what valuable new capabilities it may give the human race some detrimental consequences also will result. Fire, the wheel, rope, hammers, dams, bridges, guns, electricity, airplanes, x-ray machines, nuclear reactors—all can be misused as well as employed advantageously and can cause accidental harm in their use or manufacture. Every drug has side effects, however minor; all chemicals produced are potentially harmful; every vehicle or operation can go out of control. Every new structure or technological activity interferes with the natural environment; in the judgment of some who hold a certain sense of values, that is itself automatically a dis-benefit.

The undesirable effects of the world's technological operations at one time occupied the far background. If the free market or government controllers of a segment of the nation's activities perceived that a project promised adequate benefit, and if the necessary resources were available, the program went ahead. If we were living in such a period, or if we were to fantasize that no negatives would ever again accompany technological implementations, economic and social benefits alone would set the criteria for decision making and action. We would expect new drugs to enter the market if they accomplished some visible "good" for a sensible price; we would not worry at all about any corollary "bad." We would push ahead with nuclear reactors or synthetic fuel, considering only their comparative costs, availability of resources, and timing of need. We would forget air pollution and safety standards in design of automobiles. All rivers, lakes, and ocean bays would be permanently clean. We would not concern ourselves with toxic waste disposal or pesticides or food additives.

In this fantasy, we would no longer have to fear another existing deterrent to full use of technology: over-regulation by government in the name of safety, health, and environmental protection. If no negatives were being generated, rules to curb them would be unnecessary. The government's regulatory staffs could busy themselves amply with parallel duties in other areas, such as prices, antitrust actions, interstate commerce, and utility rates. But we are decades past the time when ignorance or pleasant dreams permit us to expect unlimited technological expansion or even unfettered continuance of all existing activities.

As an example, consider that tens of thousands of different chemicals are being manufactured at the present time and a thousand new ones are added every year. Some are produced at the rate of billions of pounds annually. Theoretical analyses have suggested that more of these substances might be hazardous than we have recognized. Over 100 billion pounds of ethylene dichloride were created in the United States before it was found to be a strong carcinogen. Vinyl chloride was being made at the rate of around 5 billion pounds a year before tests showed it to be cancer-causing. As another example, the National Academy of Sciences noted in a 1975 report that pest control practice has reached the point where a billion pounds of toxic matter is introduced annually into the environment. The report also stated that the government's knowledge of the extent of its use and its potential harm was highly inadequate. As still another illustration, there is evidence that drinking water is made carcinogenic under some circumstances by the formation of cancer-causing chemicals through the action of bacteria-killing chlorine.

A tough problem in identifying environmental carcinogens is that it may take 20 or 25 years before their effects on the human species can be noticed. This is true not only of industrial chemicals but of cigarette smoking, the lower levels of the radiation set off at Hiroshima, radium, coal mine environments, asbestos manufacture, and others. This makes it terribly difficult to know the level to which a chemical dose must be limited in order that those exposed to it can remain adequately safe. Often it is not even known whether the degree of hazard is related to the continuity or size of the dose. For some chemicals it is conceivable a threshold exists below which there is no effect.

Carcinogens in the environment are not found only in commercial chemicals. Fat in our diet is a known factor in colon and breast cancer. Charcoal-broiled steaks contain charred protein which includes mutagens. Overall, it is estimated that 80 percent of all cancer disease is of environmental origin. If this is true, then in principle most such cancer should be preventable. Government-sponsored investigations and government standards and controls are necessary. Even if some of the known harms are trivial and the benefits clearly enormous, our technological society now and forevermore is going to be a controlled society, specifically regulated in an attempt to ensure that all foreseeable detriments are held to tolerable levels.

This is a permanent policy, but in practice it is nevertheless a very ambiguous one. Defining accurately what is tolerable is essentially impossible. The unwanted ills potentially present are infinite in number and, consequently, unmeasurable in toto. Imagine that for every existing technological activity — production of food and chemicals, operation of autos, airplanes, nuclear reactors, or coal mines — we could list and measure each and every possible menace to health and habitat. We still would not know what threshold level of those dangers we should designate as acceptable. Even if we knew the effect on our bodies of everything we come in contact with in the atmosphere, water, food, and the total environment, we still could not say how big an effect we should allow.

What we define as tolerable translates to how much we are willing to risk losing. Who is to say what degree of lowering of our life expectancies or our comforts or vigor or enjoyment of natural surroundings we should be willing to countenance? When it is a matter of value judgments, in which we all differ even when we possess and agree to act on the same facts, how do we specify the threshold of harm? If we merely insist the harm be negligible, then how shall we define negligible?

To take the other extreme, no problem would exist if the technological route in life offered us no benefits. But it indeed does. Without further fantasies — like going back to a society of no wheels, tools, fire, or cloth — we cannot seriously plan on arranging a non-technological society, only one somewhat less technological, one that limits the perceived harms of technology while it seeks advantages from its continuation or further implementation. Some speak of getting rid of the automobile and returning to horses because automobiles foul the air. However, before the advent of automotive technology, when city transport depended on the lower technology represented by the carriage, the active horse population in New York City produced a few million pounds of excrement a day in liquid, solid, and gaseous form. A powerful stench was always in the air and manure dust covered clothes, windows, furniture, and faces. Extrapolating that pre-technology urban transport to today's higher city density and expressing the negatives to health quantitatively would not divulge attractive new routes for rearranging our society.

Sensible decisions as to what to allow or foster in technological change require that we identify and measure the benefits as well as the negatives. Unfortunately, for every facet of society that technology touches, the pluses generated can be as numerous and as unquantifiable as the minuses. When we know detriments can exist and we identify them, we have only the beginnings of sound laws and practices of regulation and the rendering of tradeoff decisions. A listing of rewards accomplishes no more. Both lists will be perpetually incomplete, with only some items called out and countable. Value judgments will apply to each category and people will differ as to those values. How much of a gain should we insist on before we should be willing to accept a given negative? This question is as frustrating to ponder as the opposing one: How much of a price should we be willing to pay to realize a given gain?

Knowing that the decision making required for us involves two lists, each semi-arbitrary, incomplete, and unclearly defined, we sit in the ridiculous position of trying to pit one against the other. How are we to balance the good against the bad with limited knowledge of each, no clear scale of judgment of either, and no basis for weighing one against the other?

Despite this quandary, we have created regulatory agencies in the government. All reasonable citizens in a democracy want the government to prevent harmful activities. They expect the government to decide which activities of individuals or organizations are sufficiently hazardous that it must step in to halt or alter them. Moreover, when we delegate to government agencies the power to set the standards, we know all regulation ultimately has to be a political process. Conceptions of rights, fairness, and justice will somehow eventually dominate and will go beyond purely economic considerations, cost–benefit tradeoff analyses, and science and technology facts.

To comply with regulations generally increases costs in industry, and these costs are passed on to the consumers. We all pay the bill for regulation. Very often the benefits are not assessable, at least not objectively or quantitatively in dollars and cents. We usually cannot put economic measures on the decrease of hazards and the corollary improvements in health or the prevention of accidents and deaths. The idea of a free market setting economic values is suitable for some activities of society, but not for all. There is no marketplace where an individual can buy one extra year of life or a month's supply of natural air that is healthful and pleasant to breathe. No broker can quote a price for these items. Whatever its cost to industry, regulation by government is here to stay, measurable or not, political or not.

Regulation is unsound and illogical unless it includes recognition of the tradeoffs among pluses and minuses. The public wants safety, but total or absolute safety is unattainable. There is no such thing as zero risk; a crusade that seeks it is certain to fail. A determined effort to reach it could only generate an expensive bureaucracy reaching into every nook and cranny of life and disrupting our society with no chance of succeeding. Regulation that is too severe has a negative impact on employment. It discourages new investments. It hurts our ability to compete with other nations in the world market and de- creases average incomes. People who are made poorer have im- paired health and a shorter life just as surely as do normally healthy, middle-income citizens whose environment we allow to become polluted. If the economy is hurt enough, the resulting social instabilities can lead to cruel impacts on people's lives and even to wars.

The tradeoff between improving the environment and increasing the energy supply is typical. If coal is substituted for natural gas, we might have more energy available and less chance of economic and social chaos due to energy shortages, but we might also have more environmental degradation and safety hazards. Limiting energy and hurting the economy is bad. Allowing more pollution and risking more accidents is also bad. There has to be a balance between the two negatives. The lack of a working government system to perform broad tradeoff comparisons and make balanced decisions is well illustrated by the government's past handling of the entire energy problem. First the government set a low ceiling price on natural gas, discouraging further exploration and simultaneously increasing demand. This ceiling price was stubbornly kept on even though double-digit inflation ar- rived and greatly magnified the mismatch of supply and demand. The government, in an unrelated coincidence of timing, then imposed drastic new safety and environmental controls on use of coal. In another action having no tie-in at all to the antipollution program, the government froze prices so that no one could offset cost increases to meet the government's environmental regulations. To cut air pollution, changeovers to oil and gas were mandated for utilities using coal. A little later, reacting to OPEC actions, the government required greater use of coal.

Similarly, the government introduced severe air pollution restrictions on automobiles without considering the impending oil shortage. MPG performance became lower, gasoline usage increased, and more refining capacity was called for. A shortage of unleaded gasoline then was created inadvertently because the EPA did not concern itself with the fact that its auto-exhaust emission rules would raise the demand for unleaded gasoline. When refiners take the lead out of gasoline, they must use more of the scarce natural elements in crude oil to give gasoline its necessary anti-knock properties. Less gasoline is accordingly produced from a barrel of crude in making unleaded fuel, which means that even more refinery capacity is needed. But at the same time, new restrictions were placed on refineries. While one part of the government, the EPA, pushed the demand for unleaded gasoline upward, the Department of Energy set a ceiling on the price, discouraging the expansion of capacity. Moreover, by specifically being anti- the large petroleum companies, the government further discouraged commitment of funds to build large refineries, which are more efficient than small ones in making unleaded fuel. The insistence by EPA on the rapid changeover to unleaded auto fuel appeared wise to that one isolated agency, but the overall effect, the result of the interactions, was not anyone's responsibility to assess. Meanwhile, the air is being increasingly fouled by automobiles moving around the streets in search of an open gas station, or burning fuel while in a lineup.

Being accustomed to inconsistent, illogical, and impractical decisions by government whenever it tampers with the marketplace, the citizens of the United States did not view with amazement the government policy in energy which simultaneously preached conservation, encouraged dissipation by keeping conventional fuel prices low by artificial government controls, made development of new domestic energy sources through private investment less attractive, then developed a program of government-funded assistance to pursue new energy alternatives. Nobody is in charge of the tradeoffs between having an adequate energy supply and sound standards for air pollution and safety. Federal regulation aimed at protection of safety, health, and the environment is often inconsistent and, on a common-sense basis, often unreasonable. It is too unpredictable and leads to unproductive diversion of resources, especially since regulatory agencies often mandate precisely how standards are to be met rather than allowing industry to innovate to meet their requirements.

One or another government agency handicapped the use of Alaskan North Slope oil by out-of-context concern with the cosmetic effects of drilling in uninhabited and unreachable regions. To secure approval for a California-to-Texas pipeline, the Sohio Company spent five years obtaining 700 separate permits from regulatory authorities. Then, seeing no end of legal challenges, the company gave up on the pipeline. In the present political climate, it makes less sense for oil companies to make heavy investments in oil exploration and production when they see a decreasing chance of getting the investment back in view of the government's demonstrated disinterest in return on investment.

We can learn a great deal about the shortcomings of our decision making by comparing similar potential technological developments in Canada and the United States. Canadian tar sands and American oil shale are both huge energy resources with similarities of potential benefits and drawbacks. Both sources need additional R&D to become major suppliers of a substitute for OPEC oil. Canada has moved along so as to be almost ready for commercial production. America's activities have been relatively minor. In fact, two commercial plants are in operation in Canada at costs of billions of dollars. They have involved a cooperative effort by government and private industry. Nothing comparable has happened in the United States.

Canada has shown as much interest in protecting the environment as the United States. In fact its legislation dealing with environmental matters began to be passed years before similar U.S. legislation. The Canadian authorities seem to have a mechanism for comparing reasonable environmental protection against a desire to make use of the resources. The typical U.S. approach, on the other hand, has been one of confrontation, with no one directing the overall tradeoff. Permits required in the U.S. for a single project in oil shale number in the hundreds, and failure to gain any one of these approvals is enough to halt a project or at least delay it so as to threaten it financially.

In Canada there are tax incentives dealing with depreciation. U.S. depreciation policies amount to dis-incentives. For example, in Canada depreciation may be taken during construction, before any operations have started. In fact depreciation may be charged as spent up to 100 percent against income from the project.

United States pricing policies have made no distinction between natural crude oil and synthetic oil produced from shale. This is the same as saying that shale oil, if produced, would have to be priced below OPEC oil (as is our average mandated price)* and, of course, below total cost. In Canada the product from their tar sands activities was allowed to operate at the free world prices, much higher than Canadian domestic crude oil.

Most of the tar sands in Canada and the oil shale in the United States are on government lands. The Canadian leasing policy has been liberal and realistic. The same cannot be said for leasing of U.S. land. For one thing, the leasing of federal lands in the U.S. is limited for any one company to an area of land believed to be sufficient to support one shale oil plant. Since one such plant costs a few billion dollars, and the experience gained cannot be used on a later development by the same company, the incentive to create the plant, which was risky in the first place, is greatly reduced.

Let us go to fields other than energy for examples of the unacceptable scarcity of the use of tradeoff comparisons by the United States government as bases for regulation. The Department of Agriculture has estimated that if farmers were stopped from using modern pesticides, crops would decline 30 percent and food prices would go up 75 percent. Millions of people around the world would go hungry because they depend on U.S. food exports which would no longer be available to them. But unregulated, free use of all pesticides is out of the question as too hazardous to health; their manufacture and application must be conformed to strict standards. Standards for pesticides obviously should be based not alone on the negatives of their use but also on the negatives of their non-use. One consideration should be pitted against another and a clear responsibility should be called out to cover this balancing requirement. Nobody performs that task now.

Recently a new herbicide for major grain crops was introduced. It is said to be environmentally superior to many competitive products that have been on the market for years. This novel herbicide was the result of 15 years of research and it took 5 more years before the product received official U.S. approval for use with some major crops. It is estimated that several more years will be required before it will be approved for application on other crops. This suggests that if the regulatory process could be speeded up, the environment and public health would gain rather than be harmed. Of course, an early decision is risky. It is safer to take lots of time before allowing something to enter the market. Overdo this, however, and we are denied the

*Legislation in process will remove this handicap in a few years.

benefits. We need a tradeoff of benefits versus risk. This is rarely now a factor in the regulatory or approval system.

The negatives generated by technological advance constitute the core interest of regulatory agencies such as the FDA (Food and Drug Administration), OSHA (Occupational Safety and Health Administration), and EPA (Environmental Protection Agency). These agencies are charged with regulating in the limited sense that they are expected to investigate specific operations, to set standards, to police, ban, or approve detailed activities in those fields. Their charters are narrow and they are essentially negating agencies. They exist to guard against harm and are not asked to accentuate the positives. They were not set up to study broadly what should be done in order that the nation profit most, all standpoints considered. They generally are not directed by law to compare benefits and dis-benefits of the activity under regulation, but rather to examine only the dis-benefits and direct a halt or a modification of the activity, or refuse to allow it to commence if the hazards are too high. Some of the regulatory legislation—the Clean Air Act, for instance—specifically precludes the deliberate weighing of benefits versus harms. In some instances (FDA, for example) the rules include being sure that evidence of positive benefits exists before allowing a new product on the market. However, as a whole the legislation setting up any regulating agency is weak on assigning or even hinting at tradeoff duties and balanced decision making.

Even if a regulatory agency is not directed to avoid tradeoff analysis, the political rewards and the motivations for the regulatory agencies to balance the pluses against the minuses are non-existent. The safe thing to do is say no or refuse to say yes and this causes, in actual practice, a focus on the negatives. Regulatory personnel quickly learn that permitting the public a benefit will earn no applause, but allowing a conspicuous negative, even if small, may be penalizingly embarrassing. An agency can get into trouble by approving something that later turns out to have a noticeable dis-benefit. Most particularly, no one gets a reward from Congress or is voted into the Hall of Fame for displaying the courage to make a difficult decision, one where the law is vague and the evidence is that approval would yield real gains along with some undesirable consequences.

In the listing of blame for flaws in the United States system of regulations against all forms of hazards to safety, health, and the environment, Congress must be placed at the top. The practice of creating a new agency every time a new potential damage surfaces has produced narrow thinking in the regulatory process. No really effective control exists over most of the regulatory agencies. Each new regulatory empire is always constructed to be independent of the Executive Branch of the government and is not responsive to elected officials, such as the President, who must answer to the voters. Congress rarely makes the mission of a regulatory agency clear. Congress is political, but so is the value judgment problem; thus we might hope for a happy match of natural interests. With value judgments varying greatly, as we discussed earlier, one would hope Congress would jump in and be specific. Unfortunately, Congress merely directs an agency to diminish or eliminate risks, or tells it something equally vague or impossible. These guidelines do not tell the agency whether to tolerate a trivial hazard when the cost of removing it is enormous while the banning of it will deny us a benefit which could be great (if it is, say, a highly useful drug).

The U.S. regulatory setup lacks the essential functions of comparing benefits and dis-benefits, making tradeoff analyses, and forming a balanced decision after considering all alternatives. In fact, the rules constructed by Congress to control the regulatory agencies often do not merely forbid a balanced decision but actually require an unbalanced one. The so-called Delaney Amendment to the Food and Drug Act tells the agency concerned that costs must not be considered in any way when regulations are constructed. This begs for over regulation. It says to the agency that no matter what the cost or how slight the potential benefit, its duty is to call for the most severe standards.

Congress has committees to oversee the work of regulatory agencies. Why do these committees not readily spot over-regulation when it exists, or the lack of motivation, or the responsibility for an agency to make balanced decisions instead of playing safe by disapproving or stalling approval? Agency hearings before Congress have become arenas for representatives to criticize an agency for a specific (attention-getting) boner, a failure to recognize some hazard and regulate it firmly. At a minimum, it is reasonable to expect that Congress would take an interest in the cost of regulation. A $200-billion-a-year expense for industry in meeting government regulations is comparable with the nation's capital investment. Such a sum is in the same ball park as total tax revenues from business. Congress could be excused if it were spending roughly as much time studying and justifying regulatory costs as it spends on taxes. But, of course, it doesn't. Without a budget for regulation, Congress has no systematic way of controlling the price of regulation to the nation.

We badly need a consensus to modify substantially the present regulatory system. Many regulations should be altered, diminished, or broadened to recognize the variations in practical circumstances. The situation today is one of voluminous documentation on minor issues and long litigation and delay in the approval of major projects. The regulatory system we have set up was intended to protect us, not to eliminate industrial growth or worsen vital shortages. Environmental legislation, such as the Clean Air Act, quite properly recognizes the importance of protecting against adverse effects on human health, but essentially gives no recognition to the economic dis-benefits that may accrue either as a result of the speed at which regulatory standards are evolved or the level of the standards. Consequently, conflicts remain between the desire for clean air or water and safety on the one hand, and economic growth or improved management of resources on the other.

No matter how much we might improve the effectiveness of government in the necessary tasks of discovering and assessing negatives, this will still not cover the function of making comparisons. To compare and balance the positives and negatives of any project or technology advance requires the calling out and measuring of the positives as well as the negatives. Making a decision as to a technological implementation is usually going to be difficult even if a complete listing of the good and the bad consequences of proceeding or not proceeding, of action or inaction, has been drawn competently and compared.

The difficulties of tradeoff decision making, the absence of organized responsibility to accomplish it, and the penalties of this lack can readily be illustrated by an example. The eastern region of the United States has a refinery capacity for less than a quarter of all the oil consumed in that area. No new refinery has been built on the East Coast for over 20 years. Refined petroleum products must be shipped from a distance, which increases energy dissipation for transportation and adds pollution from the dissipating of that energy. For many years, struggles have been engaged in by opposing groups: those seeking to locate a new refinery somewhere along the East Coast and those striving to prevent its being built. One proposal, in contention for many years, is construction of an oil refinery on the banks of Chesapeake Bay. (The refinery's cost is being estimated at close to one billion dollars. When originally considered, it was figured at $200 million, but the price more than quadrupled during the long fight for permits.) Some residents of the area want the refinery as a new boost to the economy. Others fear that it will jeopardize the Bay's shell fishing industry and valuable tourism income. Involved, in addition to those who would build and operate the facility, are numerous citizen groups, the Department of Energy, the Department of Interior, the National Oceanic and Atmospheric Administration, the Commerce Department, local government groups, the Army, the General Accounting Office of Congress, the Environmental Protection Agency, the Coast Guard, and others. No one of these groups has decision power; no one is charged with the responsibility for making tradeoff analyses.

We could continue for many pages with examples of shortcomings of our decision making when technological activities are involved and where both benefits and dis-benefits would result from launching or continuously preventing such activities. For instance, there is the Georges Bank project. This area stretches 50 to 200 miles off the coast of Massachusetts at a region where the Labrador Current and the Gulf Stream converge and stir up nutrients. Hundreds of species of commercial fish and shell fish spawn or feed there and the total catch over the next 20 years is predicted to be worth some $3 to $4 billion. Geologists estimate that during this same period $7 billion or more of oil and gas can be obtained from a limited portion of the area. The government is getting ready to sell petroleum leases, and strong controversy has arisen over whether the search for and extraction of petroleum products from the region will affect the commercial fishing.

From the standpoint of monetary value, the petroleum seems to have it over the fishing. However, a simple overall comparison is probably not sensible because the leases being offered are in the supposedly safest part of Georges Bank, where the currents are apt to carry any spilled oil out to sea. A bad spill would harm some species, but only those that happen to be spawning at the time and all experts say the damage would not be permanent. Oil drills and fish have been existing compatibly side-by-side in the Gulf of Mexico, the North Sea, and other waters. Nevertheless, Georges Bank is surely vulnerable to damage, and accidents could occur during the 20 years that individually might involve a $100-million loss. Many agencies of the federal government are involved, but again, the power of decision and even the ultimate responsibility for setting and policing regulatory standards are far from clear. At this time it is not certain whether petroleum operations will be permitted at Georges Bank.

Another example shows particularly how our delayed and indefinite decision making is putting us at a disadvantage in the international marketplace. California competes with Japan in buying LNG (liquified natural gas) delivered by tanker from the Pacific Basin gas reserves of South Alaska, Indonesia, Chile, Malaysia, Australia, and other locations. California grew up with gas and, because of air pollution problems, gas remains its preferred energy source for non-transportation or stationary activities. California's location makes LNG shipped in from outside a thoroughly practical approach. Seeing the future decline in America's lower states' natural gas supplies some 10 years ago, the gas companies of California commenced to arrange for LNG deliveries from sources having some two or three decades of substantial supply available. Indonesian supplies were contracted for by California gas companies starting in 1973, before the Japanese made similar contracts. However, the Japanese share of the gas started flowing in 1977. The earliest the United States now can be prepared to receive this gas will be 1983. A 10-year period will have been used up just to get approval on a terminal site for the tankers.

Similarly, the Alaskan North Slope gas pipeline project was initiated over a decade ago. The Canadian Prime Minister, the houses of Congress, and the President of the United States all blessed the project in 1977, yet it is still in a regulatory approval state. It may never be brought to fruition. A minimum of several more years, perhaps another decade, will be required, at our present pace of making decisions.

Starting with the Pure Food and Drug Act of 1906, which centered mainly on food and matters of purity and fraud, we have added in recent years regulatory laws governing therapeutic drugs, cosmetics, medical devices, occupational environments, pesticides, children's sleepwear, automotive safety equipment, nuclear emissions, and pollutants in water and the atmosphere. But all of this regulatory effort is still narrowly focused, with little comparing of benefits and dis-benefits. Technological feasibility, economic gains and costs, consideration of other alternatives, effects of specific regulations on other government programs and on overall national economic, physical, and social health are all factors which would enter the deliberations if a tradeoff responsibility accompanied government regulation. Since this mission is lacing, it has become very common in recent years for aggrieved parties to seek administrative appeal from regulations through the courts. Litigation has become so frequent that regulatory administrators' initial decisions are often rendered academic. Often the agencies themselves look forward to the step of winning in court.

An example of specific court action on a proposed regulation is the recent striking down by the U.S. Court of Appeals of a regulation by OSHA (Occupational Safety and Health Administration) concerning the handling of airborne and liquid benzene. This case is particularly helpful in illustrating that a regulatory agency usually acts merely as a negating agency and that on matters related to science and technology we are missing the function of properly comparing positives and negatives and arriving at balanced decisions.

Benzene, like gasoline, is a highly dangerous, potentially explosive chemical. If breathed in substantial concentrations, it makes one ill and if taken in over a prolonged period it may cause cancer. Regulations going back some quarter of a century originally limited the allowable molecular concentration in industrial establishments to 100 parts per million; this was lowered over the years to 10 parts per million. This hardening of the standards was based mainly on viewing benzene as a hazard because it is highly flammable. Cancer-causing effects had not yet been recognized during that period. Even at the present time, very limited data are available on the maximum exposure times and concentrations allowable if cancer inducement is to be avoided.

A recent OSHA edict decreased the molecular concentration by a factor of 10, down to one part per million. As the Court observed, OSHA had not performed any tests to learn what improvement in protection, if any, would result from such reduction. Industry admittedly did not know that answer either. Would adhering to these more severe standards save 100 lives annually? One life? Benzene is present all about us (if we are to consider occasional molecules as a presence) because it is put into the air by trees, petroleum, coal, and various petrochemicals. But nobody knows how large a concentration can be tolerated without affecting the body. We don't know what threshold level should be set for benzene.

On the other hand, it was quite readily apparent that if OSHA's new proposed standards were adopted, it would cost the industry over $500 million. Immediately, then, a tradeoff question arose: some clearly disadvantageous, large economic penalties versus some unclear, possible, but perhaps totally absent, health benefits. Who should decide this question? OSHA thought of itself as presumably having the duty to do so, but its charter does not make this certain. At any rate, OSHA simply was going on the assumption that if restricting benzene content in the air to a very low value is a good idea, then reducing it to a still lower value must be a better idea. They did not think of themselves as having to pit the cost against the benefits. The $500-million cost was not their business, they assumed. Indeed, they could have taken the view that even if one life were saved it might be worth any amount of money, not only $500 million but perhaps $500 billion. But surely some cost is prohibitive and some of the expected benefits must be ascertainable. At least, this is what the judge decided when OSHA was taken to court by the industry. A Circuit Court of Appeals ruled that OSHA could not apply the more severe standards.

As this book is being written, the case is being appealed to the Supreme Court.* Is it not an unacceptable way to run the country if the Supreme Court has to settle whether the permissible number of molecules of benzene in the air should be one in a million or ten in a million? Surely we should be able to arrange a better system to handle such tradeoff decisions. The courts must interpret the law and ensure justice and fairness, but legislation has not furnished the United States with a practical system of arriving at balanced judgments on the positives versus the negatives of technologically based issues. With this function missing, we can easily get over-regulation. We also can evoke some very peculiar government actions in which the opposite of the end result being sought may occur.

Take the example of the current clean air "offset" requirement, which mandates that "old pollution" has to be cut down before new plants can be built in the same geographic area. Since it is not always practical to make an old plant pollute less, this rule means that new, efficient, low-pollution plants are discriminated against in an established area while existing plants that are polluting heavily are allowed to remain and go on doing so under a grandfather clause. The effect is to penalize the economies of existing cities and industrialized areas. It also discourages investment in up-to-date technologies that underlie the new cleaner plants. Consider that if to build a new plant also requires building a whole new industrial area, moving people and creating infrastructure around them—along with the usual investment risk in new technology—the overall reward–risk ratio can become too low to justify the investment.

Note also what is happening in the therapeutic drug industry where research on drugs (not to be confused with producing and putting the drugs on the market) is now being regulated. Two decades ago it took five years and $1 million on the average to work a new drug product through the maze of the government's regulatory apparatus. Today the cost figure has been estimated at $20 million and the time is 10 years. In this period the rate of new drug introductions by American firms has fallen by 50 percent. Has our innovation process been slowed by over-regulation or by a misguided goal to build a risk-free society, or are we learning to really curb the harms of scientific and technological advance in drugs? Should we be glad

*Since this was written the Supreme Court has ruled against OSHA.

the tradeoff is as it is and thankfully pay the price in time and money, to prevent the dis-benefits of hasty introductions of bad drugs? What are we missing in benefits that the new drugs might bring to the nation's health? We don't know. We don't know how these tradeoffs work out and no group has the definite function of trying to find the answers to these questions.

Early drug research has a very safe record; there has never been a death or serious injury from the performing of drug research. To be sure, drug disasters have occurred, but with marketed drugs, not those in the research phase. It is reasonable to question, therefore, how much added safety the government obtains by regulating research in drugs. Regulation in the research phase causes a substantial burden to fall on innovators in pharmaceutical drugs in the United States because they must supply information that can only be obtained by costly investigations. Left to their own devices, the innovators would first do initial experimenting, to decide whether the drug might have real promise. Only if they conclude it does would the drug then deserve and receive the major investment required to bring it out and defend it before the government's regulatory agency. In effect then, regulating the research inhibits research, which in the end tends to deny the potential benefits of new drugs to the citizenry.

In the decade of the 1970s, American drug companies increased their annual R&D budgets in foreign countries from under $50 million to over $250 million. A large part of the money spent overseas goes into trials in which clinical pharmacologists in other countries give experimental drugs to groups of patients and healthy volunteers. Aside from whether we think what they are doing is right, other countries have different views of the tradeoffs between the dangers of new drugs and the values they might provide. Their views are more conducive to research than ours. If pharmaceutical R&D moves abroad, foreign countries, not we, will be penalized by the hazards. However, they also will be the early beneficiaries of the breakthroughs and the financial returns. Perhaps it has worked out that our approach to regulation of drugs has given us high protection, while the delays and disapprovals have cost us little in unrealized gains. If so, that would be in part an accident because no one is really in charge of making such tradeoff analyses. The regulators limit themselves to a narrow decision to negate if necessary and approve, if they do, without urgency, in contrast with comparing the good with the bad and attempting a balance.

What organizational steps should be taken toward solving the problem of the missing tradeoff analysis and decision function? One would be to alter the present government regulatory agencies dealing with safety, health, and the environment so that they are regarded henceforth exclusively as investigative agencies specializing in the negatives — definitely not decision-making groups. Then we should provide for the necessary tradeoff comparison and decision function in another way. We should relieve the investigatory groups of their vague aura of responsibility to consider positives as well as negatives and to reach balanced overall decisions. When it comes to clean air and water, nuclear safety, health hazards from toxic chemicals, occupational health and safety, and purity in food or drugs, the agencies watching these areas should be equipped with the required tools and budgets to track down, understand, and estimate all the detriments of proposed or existing activities. Their work should finish with their presenting their results and making recommendations, particularly when called upon to do so because decisions need to be made.

The operations of the Federal Bureau of Investigation constitute a useful parallel and guide. The FBI investigates crime and tracks down criminals. It does not try them and sentence them. It does not decide on the place of capital punishment, or whether jails are to punish or rehabilitate. It is properly named an investigatory agency. When it finds a culprit it turns over the prisoner, and the evidence it has found in its investigations, to another part of the government, which takes the case from there. It would be patently ridiculous for the FBI to have its own court, judge, and jury and then internally ponder the case and decide what to do about the criminal. Equally, it would be nonsense for the FBI to be the end of the line, busily investigating, with no place to take its results for the obviously required next steps of acting on the data and arriving at decisions about the case.

Presumably some entity always exists, private or public, such as a drug manufacturing firm or an electric utility, that wants to move forward with a product or a project (either because the firm doesn't understand the negatives well enough to be deterred or believes they are outweighed by the positives). We will have then two clearly opposing parties, each with unambiguous missions and desires, one dominated by the advantages of some activity it wishes to carry on, the other, the government investigating agency, ready to say what detriments the proposed activity will bring.

Perhaps the two interested and contesting parties will agree the activity ought to start or continue, or conversely, should halt or not start until the dis-benefits called out can be lowered. If they do not agree, the decision of a moderating board of some kind is needed to settle the issue. This decision board, unlike the negating or investigatory agency, should have the role of balancing the good against the bad in the interests of the nation as a whole. The board should have the unquestioned responsibility for banning or stopping, or approving the commencing or continuing of, proposed technological operations. It should expect to act only when a ban or set of standards to be adhered to is recommended by the government investigatory agency covering that field and, if at the same time, that regulatory recommendation is contested. Such contesting would presumably be by those who wish to go ahead with or continue the activity the investigatory agency suggests should be closed down, prevented, or amended.

Perhaps the two opposing parties, the would-be operator and the government investigator, most often would settle their differences by negotiation "out of court." But if they do not, they would come before the proposed decision board to present their cases. The pluses and minuses of the activity would then be brought out as clearly and thoroughly as possible through the board's hearings. Then the board would render its decision. The decision board would be charged with the requirement of objectivity and breadth, with no limit as to the scope of its deliberations as it ponders the specific case before it — except that it must act within the laws of the land.

As the boards go about their job of rendering decisions, surely the nation's courts of law will be sought out from time to time by one or another interested party in an effort to change or force a decision by one of the new boards. If the legislation setting up the boards is competently written, its decisions will be interfered with by the judicial branch of government only when the boards, in their decision making, have violated the law, overstepped their charter, ignored other legislation whose conditions are binding, or failed to consider the rights of a minority. For example, the board might be correct in halting rail shipments of certain dangerous material unless certain safety standards are met, but the board might be wrong if it directs that a particular railroad must provide the transport under the stated requirements. If the railroad prefers to pass up the business as involving too much financial loss, especially if severe standards have to be met, it would be totally within its rights. As another example, the board's decision may be the best for a vast majority of citizens but yet be unacceptably penalizing to a small group.*

Our courts already sometimes sit on technology-based cases involving benefits versus detriments. The Supreme Court now has before

---

*New Mexico is the source of 50 percent of the U.S. uranium supply. This fuel is needed by the nation but acquiring it is hazardous in the extreme to the Navajos of the region. (See editorial by Alan Ramo, Los Angeles Times, Sunday, June 1, 1980)

it some dozen cases involving a challenge to government regulation. No matter how we might modify national decision making as to technological applications, the Constitution and the courts will always be with us and decisions made in any way may end up in the courts. But we can do better than guaranteeing that litigation will be required to settle every issue. Properly designed, the decision boards should be able to resolve most situations.

Once we have accepted the value of a clear government agency role to investigate negatives, harms, hazards, and dis-benefits either existing now or potentially present and damaging in the future, it is sensible to consider bringing all such duties together in one agency. All dangers to the human body and the land, atmosphere, drinking water, rivers, and seas from drugs, radiation, chemicals, fire, explosions, collisions, and all other sources require the same kind of ensemble of scientific and engineering talent, measurement laboratories, and statistical analysis. They all require that expertise in chemistry, physics, biology, physiology, and toxicology be organized into investigatory teams continuously handling projects and questions that will come and go as various technological activities of the nation develop and mature and are terminated or altered. For instance, food involves a chain of events, each with its separate dangers, from the watering and fertilizing of plantings to their protection through growth, processing, storage, and delivery. A thoroughly competent investigatory agency, to cover this array of disciplines, must be well equipped with experts and laboratories covering diverse fields. Similarly, assessing hazards in the water, air, and chemicals used by individuals and industry will usually require considerations now handled by a number of separate regulatory agencies. As further examples, automobile and airplane accidents include fire hazards, which relate to fuels and materials used. Fuels, whose combustion in automobiles pollutes the air, may also harm workers during production. All hazards require similar considerations over a broad spectrum of science and engineering.

A single investigatory agency for Safety, Health and Environmental Protection—let's call it SHEP for short—if set up to cover all hazards, would have to be organized into many project and functional units. For instance, a chemical analysis laboratory might assist everyone but would not house all chemists because some would be on teams specializing in industrial chemical processes, or syn-fuels, or pharmaceuticals, or fertilizers, or pesticides. But there would be advantageous flexibility of organization and unity of mission if we created one very strong government unit seeking always to be expert on all the negatives of all technological activities. No need would exist, as now, for Congress continually to discover new areas — occupational safety, automotive safety, environmental protection, drugs, etc. — and then create still another new government agency to handle it.

Let us elaborate now on the proposed decision boards to fill the missing function, the tradeoff balancing and decision-making responsibility. While a single government agency, as just described, may be the way to handle the investigatory duties of government on hazards to life and protection of the environment, we should not create a single new super-agency to cover the decision function for every area of the nation's technological endeavors. This is because great differences exist in tradeoff analysis and decision deliberations among the many different kinds and sizes of projects, from chemical waste to auto safety, from nuclear energy to urban transportation, from depolluting of a lake to approvals for new drugs. This suggests tailoring the decision making, as well as setting the proper roles for government and free enterprise, flexibly and differently for the various separate areas of importance. We cannot solve the entire national society–technology interaction with one sweeping decision board. There should be a decision board that handles nuclear matters, another on foods and drugs, another on occupational safety and health, one on environmental pollution (air, land, water), and a limited number of others.

Should these decision boards be regarded as extensions of the executive or the legislative branch of government, that is, of the President or of the Congress? The executive is the right branch, a conclusion we arrive at by paying close attention to the fundamentals of any decision board's missions. We specifically want those boards to: (a) make, or cause to be made and presented to them, tradeoff comparisons on the alternatives of the issues brought before them, seeking a proper balance of benefits and dis-benefits on behalf of the nation; (b) integrate their sense of values to form criteria for judging the good versus the bad of each option and use these value judgments for decision making; (c) constitute a credible representation of the goals and views of a majority of the electorate. The President should appoint the boards, naming citizens of outstanding competence and character, with the consent of the Senate for each individual appointment. This appointment process will cause the boards to be adequately responsive to the country's political preferences, thus taking care of item (c). By the Congressional legislation setting up each board, and by other pertinent legislation affecting the issues the boards will ponder, item (b) will be met. By operating competently, the boards will satisfy item (a).

The decisions of the boards we propose, and the boards themselves, would be inherently political. They are not economic or technological, although economic and technological considerations would enter into the boards' pondering of issues. They are political because they depend in the end on value judgments of the citizenry. The boards, appointed by the President who is elected by the people, will constitute a pragmatically effective, decentralized microcosm of the integrated value judgments of the voters.

The need for these boards and SHEP, the investigatory agency we described, can be summarized well by quoting the words of John G. Kemeny, President of Dartmouth College and Chairman of the Presidential Commission on the accident at Three-Mile Island:*

> Our decision-making process is breaking down. The problem is whether our current political process can handle the complex issues of modern society—highly technical questions of science and technology that also involve value judgments.
>
> A good example is clearly the energy problem. There are weird discussions now, with people on both sides who have only minimal technical competence and are talking from highly emotional points of view. Emotionally, for instance, I love the idea of solar power. But I have absolutely no competence to say what it would cost, or how long it would take to get solar power on a large scale, or what environmental dangers it might pose. And I don't see how you can get an intelligent political discussion until these issues are resolved.
>
> The question is easy; the answer is not. I am still a believer in democracy, but I think some changes will have to happen in the practice of it. We have to have a forum for effective discussion of highly technological issues, so that there is a clear consensus on what science and technology say about an issue. Then the political process can make the value judgment.

Obviously, Congress can set all manner of conditions that will restrict the scope of any one board's decisions. Illogical or unfortunate as it may appear on occasion, Congress may pass laws that specify precisely how many molecules of some noxious gas may be permitted in the air as an upper limit. Then the decision board will not have

*Newsweek, November 19, 1979.

the power to allow some operation to exceed such specific standards. Similarly, Congress may ban a substance or outlaw a particular activity in a certain geographic area and the boards then will have to do their decision making in accord with these legislated rules. As minimum legislative actions, however, the Congress must alter existing regulatory agencies. Whether, as suggested here, they are lumped into one investigatory agency (SHEP) or not, Congress must narrow the duties of these agencies to the investigation of negatives and the issuance of challenges and recommendations. It then should set up the new decision boards and give them alone the power of decision making.

In the following chapters we shall illustrate how the investigative and decision functions of government can best be handled by considering energy, urban transportation, environmental quality, and other problems. We shall start with an area involving today the greatest of controversy, a field of the most demanding urgency for that kind of government decision making that balances needs and benefits against detriments. This field is nuclear energy.

7

the nuclear energy stalemate

The reason the nation commenced the spending of billions of dollars a few decades ago to develop the technology of controlled nuclear fission was to provide an efficient and clean source of electrical energy. True, some quiet predicting had gone on at that time, suggesting that readily available petroleum would start to dwindle by the year 2000 or so. But concern over a future lack of oil was not the determining factor. More important was that nuclear energy appeared to have economic advantage. Coal, though the United States' supply was known to be enough for centuries, already had lost out to low-priced and highly plentiful petroleum-based fuel. Moreover, coal was seen as hazardous to mine and dirtier in burning than natural gas and oil. The nuclear reactor, however, was judged as even better than petroleum combustion for producing electric power because it appeared to be both clean and relatively cheap—once the big initial R&D expenses are covered by the government. The power utilities, accustomed to making detailed, professional estimates of capital requirements, fuel costs, and operating expenses, concluded they could gain economically by going nuclear for most, if not all, of their needed expansion. The majority of large electric utilities made plans to do so and some went all the way to full implementations.

Safety requirements were recognized to be severe for atomic power plants from the beginning. Architectural and engineering designs thus called for unusual strength, shielding and tightness in all the external surrounding structures and the internal apparatus, plumbing, and reservoirs. Multiple indicators and mechanisms for start-up control and shutdowns were specified. Overlapping fail-safe procedures, with backups for instruments and other critical hardware items, were provided to guard against theoretically possible severe accidents, although these were envisaged as extremely improbable.

These developments brought a new fuel, uranium, into service. When some decades later adequate uranium would become more difficult and less economical to acquire, it was planned the breeder reactor would be ready to join in. Since it produces more nuclear fuel from the available uranium even as power is being generated, the breeder was seen as able to stretch out the fuel supply from decades to centuries.

Now, rather suddenly, this perception of nuclear energy is out of date. A greater urgency exists today to provide alternatives to petroleum and, from this standpoint alone, accelerating the contribution of the nuclear option would appear to possess a higher priority. However, the concern about safety has changed even more and the true costs of safe nuclear energy appear much greater than were estimated earlier. No longer is it assumed universally that nuclear power plant engineers have taken safety into account adequately and that the overall design of present reactors is acceptably free of extreme dangers. Instead, many believe nuclear reactors are inherently too hazardous to be operated, or that, if they are ever to become acceptably safe, it will be only after a vastly greater effort has been expended on every facet from location and control of the power generators to disposal of nuclear wastes.

Generating energy by nuclear fission reactions, with uranium as the fuel, leads to a number of problems not experienced when energy is obtained by other means. One is danger inherent in mining the uranium ore. Another is the accumulation of long-lived radioactive waste and the need for storing such dangerous materials with safety. Still another is that in a large-scale nuclear power plant much radioactive matter is present; in a hypothetical accident in which this matter might be well dispersed, the lives of people could be endangered over a substantial geographic area. Finally, if one postulates great numbers of nuclear power stations throughout the world, along with plants to reprocess spent fuel and breeder reactors to provide enriched fuel, absolute control of the use of fissionable matter becomes difficult. Specifically, the main nuclear bomb material, plutonium, would become available in many nations which also are capable of acquiring the know-how to produce nuclear weapons. The explosion recently of a nuclear bomb in India is evidence of this.

In the present world situation, it is not inviting to enhance the probability of further bomb proliferation.

No living person (Einstein having passed away) is recognized today by everyone as an objective expert. Indeed, highly trained individuals who have devoted their entire careers to the nuclear field are seen to disagree on occasion about some aspects of nuclear energy. For every danger-related phenomenon in nuclear generation of electric power — the probability of a catastrophic accident, the long-term effect on the population of minute amounts of radioactivity leaked to the atmosphere or the ground, the health hazards in mining and processing uranium ore, the consequences of mishaps during transportation of radioactive materials, difficulties with waste disposal — it is possible to find individuals, seemingly authoritative, who speak on opposing sides of the issue. Those who in common know intimately the details of nuclear energy generation can and do have different value judgments about risk taking with nuclear reactors and about how safe we must strive to make them as a balanced component of the overall risks of living. This contributes to the confusion. Also, for some matters not enough facts are yet known. For instance, all experts will agree that large quantities of radiation are harmful, but it is not known for sure what the effect of tiny amounts may be, so there is room here for the experts to disagree in their speculations.

Not only the professionals, but all the rest of us as well, are caught up in this matter of differences in value judgments. Since no aspect of staying alive can ever be free of risk, what do we even mean by safety of nuclear reactors? Individually, are we better off to be optimists or pessimists? How do we rate specific gambles? We know we kill and maim hundreds of thousands every year in automobiles but who argues the automobile should be outlawed? Airplanes sometimes crash. Cigarette smokers too often get lung cancer. Millions of people ruin their hearts and livers by over-drinking. But we do not ban flying, smoking, or drinking. A train carrying chlorine recently derailed in Florida and 5,000 people had to be evacuated to save their lives. Still, such trains are not being banned. Occasionally man-made dams fail and cities are wiped out, but dam building has not been made illegal and no one is demonstrating to demand it. During the winter of 1978, on a single weekend in New England some 50 to 100 roofs collapsed from snow overload. This included a coliseum and an auditorium. Had they been occupied at the time, tens of thousands of people could have died. Few have suggested abandoning these kinds of structures as too dangerous.

We are a people accustomed to taking many, many risks. Those of cars, smoking, airplanes, or alcohol are familiar and apparently accepted, if not acceptable. Nuclear power first came conspicuously to our attention as the most destructive military weapon in history. The nuclear hazard is frightening and it is new. The newly scary is always more so than a danger to which we have become accustomed. People do not know how they should react to the probability figures they hear, and are confused by the opposing arguments about the actual degree of risk. It would be helpful in working out of this dilemma-filled controversy on nuclear reactor safety if all actions and expressions of points of view, and all investigations that are carried on, would involve unfrenzied deliberateness, objectivity, a bias to act with restraint as to conclusions when faced with uncertainties, and an ability to balance opposing aspects of the issue, seeking to understand the tradeoffs. It seems that at the early stages of all confrontations of safety versus need — not just those of nuclear energy — the earliest players on the stage are precisely those individuals or groups who lack these tendencies. But that is life and that is the way the environment for nuclear energy now is and probably will remain for some time.

With all the muddle, nuclear energy is today a first-tier political problem. Politicians now feel compelled to ponder which stand to take, for or against nuclear energy, to gain the largest following among the voters. They can even do such deliberate calculating without a guilty conscience about being opportunistic instead of standing for what is best, because they realize they do not know what is good for the society and sincerely doubt anyone else does either.

Under these circumstances, to travel a totally private-sector course to a nuclear energy capability for the United States is not a real-world option. Those who argue that the free market should be allowed to solve any problem, or at least every aspect of the energy dilemma, talk nonsense when they include the nuclear energy dimension. The political disorder surrounding nuclear reactors and the differences in value judgments among intelligent people about what degree of risk exists and what chances it makes sense to take are not going to go away. No group in charge of truly discretionary private capital investment is going to rush to put its chips on nuclear power.

Even if a power utility is totally committed to the nuclear route,*

---

*In discussing nuclear energy here it is intended the area of discussion be narrowed to the generation of electric power by fission reactions. More particularly, we refer first to the so-called conventional, light water reactor (the Three Mile Island reactor is one) of viewing it as the best alternative to supply its customers with needed electric power, it cannot move forward in this chaotic atmosphere. The company would first need hard-to-get approvals from various agencies of government. If it eventually received these permits, it then could expect and would have to win numerous court actions challenging its plans. Even if all specific privileges to proceed are granted, the situation could change in a year, or two, or five, because the government might pass new rules within which it would become impractical to complete the installation. Should the plant finally go into operation, its continued operation might be enjoined by the efforts of still other groups through later court actions. These added burdens and delays placed on the utility during the life of the project would greatly enlarge the size of the required investment and lower the return on it.

In nuclear energy, the United States has reached a stalemate. A wide array of non-technological factors is intermixed with the science and technology, with no satisfactory way to resolve the issues, assessing the good against the bad, the needs against the concomitant risks. America today has no practical, working mechanism for matching nuclear technology to economic, political, and social objectives.

One way out is to abandon the nuclear option entirely as too dangerous and not needed. Is nuclear energy necessary for the United States? Must we proceed with this energy source, taking on the risks of highly serious accidents, however low the probability of their occurrence, and the social confrontations and political disturbances certain to be launched by those who believe the risk is unacceptably high? Can we not handle our energy requirements in other ways and redirect the huge technological and financial resources which nuclear energy will require toward the other options? Many billions of dollars have been spent on this controversial approach, but because we made the investment it does not follow that we must put even more behind it.

To answer these questions, we need first to agree that a decreasing availability of petroleum and a substantially increased price for it

---

which around 200 are in use in the world today, with over 70 in the United States. In these installations the heat of the fission reaction is imparted to water, making steam to turn a conventional steam turbine which drives the usual electric power generator. We also mean to include the later addition of breeder reactors to such nuclear facilities. Without the breeders, power companies would be embarking on the nuclear trip toward the dead end of an insufficient supply of reasonably priced uranium a mere several decades down the road.

should be assumed. Further, American dependence on foreign oil will become increasingly unacceptable as a source for our energy requirements, both politically and economically. Even with the highest success in conservation and exceptionally good luck in R&D on new approaches to energy, the state of the nation will continue to be highly affected by the threat of energy inadequacy. In electric power generation specifically, blackouts and brownouts probably will become frequent as will limitations on the use of electricity in homes, offices, and factories and for urban transportation and lighting. For any form of electric power generation, so many handicaps have already occurred—new site location delays, insufficient raising of capital, inability to meet stringent environmental pollution regulations—that the electrical utilities are behind in preparing even minimal expansions to satisfy the demand seen ahead.

Meanwhile, the factors influencing the practicalities of alternatives to nuclear energy are many. We cannot assume that the production of synthetic gaseous and liquid fuels from coal or shale will leap forward during the next decade because enormous environmental problems exist and very difficult arrangement making has to be accomplished with all the semiautonomous entities involved in such programs. The increased use of coal as an alternative to nuclear energy could greatly increase deaths from black lung disease and accidents in mining, transporting, and burning the coal. Mining and shipping, even at today's low level of coal use, kill hundreds each year. Acid rain from use of coal (stemming from the sulphur dioxides produced in the burning of it) is being regarded ever more seriously as a harm to the environment on a large scale. Some scientists also expect impairment of the ozone layer. There is great concern over the greenhouse effect resulting from a higher concentration of carbon dioxide in the upper atmosphere due to the burning of fossil fuel. This phenomenon is predicted to have such a negative effect on health and safety in a few decades that it will equal dozens of nuclear accidents. (We shall go into this more in the next chapter.)

Synthetic fuel from coal or shale may take off and it may not. We have to express doubt in the same way about solar energy. Discoveries may be forthcoming to make it sensible to go all-out in solar energy and cause it to provide much more than a mere few percent of the nation's needed energy supply in a decade or two; but again, those breakthroughs may not happen no matter how hard we push for them. Some say it is still possible that large quantities of oil and natural gas can be found around the world and that this could vastly change the energy supply situation. Others who have devoted their lives to professional effort in these matters say this is dreaming. Biomass may be the answer, some claim. Others with equally credible credentials say biomass will be only a modest, incremental source at best.

We do not know for sure how everything will break. Considering our record, certainly it would be foolish to assume that clear goal setting, objective investigation of options, and timely making of decisions about energy will all happily become our national pattern. It is more likely we shall continue to piddle with conservation and all alternate energy approaches. As to the nuclear dimension, the nation will probably remain divided as to its necessity and importance. Most of the public may gradually settle down to reluctant acceptance of nuclear energy as a permanent, if vaguely allotted, part of the nation's energy sources. At the same time, widespread fear of nuclear hazards will also be a long-time phenomenon and many Americans will remain unalterably opposed to the nuclear approach.

As to the first group, those who accept nuclear power, we must note that some parts of the country already are highly dependent on existing nuclear installations for the operation of their industry, homes, schools, hospitals, and the rest. Nuclear energy now supplies about 13 percent of the nation's electricity. President Carter has called for the atom's share to increase to over 35 percent by the end of the century. The country now has 72 operating nuclear plants, 94 others are under construction, and 40 more are on order. Suppose the plants operating today were all shut down in response to fears and the opposition to their continued operation. Within a matter of hours, far louder cries would be heard from those who would be prevented from continuing their daily lifestyle because of the stoppage of electric power flow. No matter how skillfully antinuclear groups might argue those plants should not have been built, or are too unsafe to be continued in operation, those critics will likely be shouted down by the users if a sudden and severe reduction of electricity supply were to occur.

To close even a few existing plants would trigger serious power shortages in New England and some parts of the South and in the Chicago area where 50 percent of the electricity comes from the atom. Nearly $100 billion has been invested in nuclear plants now in existence and $50 billion more in plants not yet completed. Some utilities have used nuclear reactors for over 20 years and no accidental deaths have been recorded due to the operations. To close down facilities now operating or to halt construction of new plants would mean economic ruin for most utilities directly involved or an outrageous and completely unacceptable increase in electricity rates. A Gallup poll taken after the Three Mile Island accident showed that the public, by a 56-to-31 percentage margin, still believed nuclear energy's risk to be less penalizing than the energy shortage that would result if plants were closed. At the end of 1979, the Congress, by a wide majority, rejected a bill to halt all new licensing of nuclear installations.

After the Three Mile Island accident, the nuclear plants at Indian Point on the Hudson began to be the subject of high concern. With Indian Point only 35 miles from the center of Manhattan, millions of people theoretically could be caught in a catastrophe. Clamors for a shutdown of the plant were largely stilled, however, when: First, Con Edison said shutdown would force it to turn to fossil fuels, the region would have to pay over a billion dollars in added bills, and severe cutbacks in electric power would be felt in less than five years. Next, others stated that the added deaths attributable to burning more fossil fuels would be certain while the chances of a large-scale nuclear accident are very, very small. Then the Nuclear Regulatory Commission announced that for a mere $100 million new safety devices (''core catchers'' to contain a melting core, additional cooling water systems, venting systems to reduce explosive pressures while filtering out radioactive exhaust) would decrease probabilities of accidents even further. Finally, the New York Times came out editorially for running the slight risk of serious accidents rather than paying the certain penalties of closing down the Indian Point reactors.

The electric utilities will continue to analyze their options and most will continue to decide that nuclear energy is a satisfactory direction in which to move — if the extreme financial and time penalties of delayed decision making can be reduced. Meanwhile, the worsening political and economic situation resulting from our high imports of petroleum will put increasing force behind the idea of not using oil and gas for electric power generation.

The practical attainment of adequate electric power without dependence on oil, gas, or coal would stimulate new ways to use electricity as a substitute energy source. For instance, electric battery-operated automobiles would be more attractive if electrical energy from safe nuclear reactors were relatively abundant. Our imports from OPEC nations, now so essential to our gasoline-driven cars, could then be greatly cut. This prospect would enhance technological effort to invent superior electric batteries. We cannot power airplanes through a plug into an electric outlet on the ground or by loading the wings with fully charged electric batteries. On the other hand, if plentiful electric power were available from sources other than fossil fuels, this could supply urban transportation systems, interurban trains, home heating, and even some chemical industrial processes as well as automobile vehicles.

Consider for a moment the other extreme. If a calamitous accident (such as some thought was happening at Three Mile Island) were to occur with any existing reactor, nuclear energy in the United States would doubtlessly be stopped for decades, if not forever. Barring this, we had better be prepared for nuclear energy's being a permanent member of the nation's energy source team, rather than its being banned. The question will be how much to rely on and expand the nuclear potential and how to go about making these decisions. If the nuclear option is here to stay, somehow we are going to have to learn how to describe and judge the risks of nuclear energy, to allow the decision making to proceed. Our indecisiveness about nuclear energy involves several factors. One is the lack of means for objective balancing of risk taking against need. Another is the need for setting appropriate roles for the government and the private sector. As to the first, the permanence of concerns about safety of nuclear reactors is paramount, and this deserves more discussion.

Reactor designers are unanimous in telling us that the basic phenomena of reactors preclude some of the cataclysmic accidents frequently imagined by critics of nuclear energy. Nuclear technology came onto the world stage suddenly in the form of a bomb a million times more powerful than any explosive previously known to man. Even though the scientific activity inside a power reactor is different from the physical phenomena employed in a bomb, and entails only the generation of heat and ancillary radiation effects, most do not understand this and think a reactor can turn into a bomb. It can't.

As to other major accidents that in principle could happen, most serious scholars of the possibilities say the odds against the required combination of events all happening at once and not being interfered with by systems that can be designed to stop such a chain of events are enormous. (Some who oppose the nuclear approach are not interested in theoretical probabilities, such as that various detrimental effects in nuclear reactors have only one chance in a billion of occurring in a century. They want no nuclear activity of any kind on this earth.) By now, some with credentials as experts have told us that nothing physically (not to be confused with psychologically) happened at Three Mile Island that was dangerous to the public. No employees were hurt or killed and no citizen on the outside was injured, they claim. There was a lot of damage to the plant through a sequence of human errors; but, despite the factors that gave rise to the accident, the plant's safety systems worked and the public was protected. The expensive repairs now required could have been averted by more investment in computer-surveillance systems to alert operators that valves were in the wrong position and by better training of personnel. There seems to have been a release of a small amount of radioactivity from the nuclear plant which these same analysts assure us was below the danger level.

Months after the Three Mile Island accident the broken reactor is still a problem for the scientists and engineers, but it would not be right to call it a mystery. Thousands of gallons of radioactive water have to be disposed of and the walls and equipment in the containment have to be cleaned to reduce the levels of radiation so that people can stay in the building to get the repair work done. The Nuclear Regulatory Commission has ordered that the cleanup should not begin until an environmental assessment can be made and the general public has had an opportunity to comment. This includes the question of how dangerous might be the disposal of the cleaned water into the nearby Susquehanna River. The plant, like others situated near large bodies of water, has dumped cleansed water before. But now, of course, there is much greater sensitivity to the issue and a good deal of scientific data collection and experimenting probably will be required before this dumping will be permitted.

The Three Mile Island accident is a reminder of the fact that nuclear reactors cannot be operated, designed, and maintained in the same way we assemble automobiles or run the Department of Energy. Admiral Rickover, for nuclear submarine activities, has created an unusual core of individuals with standards of discipline, competence, and alertness well beyond the ordinary. Surely the utilities, if they have not already done so, can do the same. If that means higher costs, so be it. Proper publicizing of a truly severe and rigid program of safety control might inspire confidence by the citizenry that the requisite safety can be made inherent in the design and operation of nuclear reactors and that it indeed exists.

The experts we have been referring to in the last few paragraphs are individuals whose professional specialties are the design of nuclear reactors and safety systems and the analysis of nuclear phenomena pertinent to these systems. We are justified in doubting the overall objectivity of these specialists, in view of their personal involvement and their natural desire for acceptance of their life's work. We also observe that only a rare few of those with professional qualifications and detailed experience with nuclear reactors ever express much doubt as to their safety. The most frequently stated concerns regarding the safety of nuclear reactors are expressed by people without professional background specific to the detailed subject matter. Included are individuals in totally remote fields such as entertainment, who, however sincere may be their worries, hardly inspire public confidence in their competence to judge either whether the nuclear approach is safe or what its relative safety is compared with the other options for energy supply. Some of the most vocal individuals are clearly seeking personal publicity, trying to create a following for a political career, or using the disarray to achieve independent social goals that interest them.

All citizens are entitled to their own value judgments. Many have formed deeply rooted conceptions, right or wrong, of the relative safeties and economic realities of nuclear and other alternatives for providing the minimum energy requirements of the nation. Some are handicapped in that they have not yet grasped that safety is relative and that there is no such thing as an approach to the energy problem (or, for that matter, any other decision in life) which is without risk. Perhaps most citizens, vocal or not, appreciate that it is not the total elimination of risks but rather the common-sense balancing of risks that is paramount. But even these individuals, because of variations in their understanding of what the odds are, will find the nuclear energy question a puzzling one for the foreseeable future.

A diversion might be helpful here, an imaginary event a century ago related to the introduction of the automobile to society. Picture, if you will, that back then some folks gathered to contemplate the possibility that, if they did not act to stop it, the United States in a hundred years might attain a population of 200 million automobiles and trucks, each depending upon controlled explosions of a highly flammable substance, gasoline, in engines inside the individual vehicles. One of the group, not to alarm the rest but merely trying to be sound and analytical, might have pointed out that this astoundingly dangerous condition would mean the country would be loaded at all times with vast quantities of gasoline in a national crisscrossing of pipelines, storage tanks, and refineries. He might have gone on to estimate that 200 million vehicles would imply billions of gallons of gasoline stored in containers or in transit between them. He might have added that each person owning an automobile would have to be willing to keep ten gallons in a tank right in the vehicle, which he would have to park by his house or, worse, in a room attached to it.

The group might have judged this future projection to be as dreadful as the corollary phenomenon to be expected: millions of vehicles facing each other in on-coming traffic at 25 or 50 miles an hour or more — surely creating collisions that would kill thousands every week! As the discussion proceeded, someone else might have observed that the mere dropping of a cigarette could cause an explosion in a car in a gasoline station. This could blow up the gas station because of its immediately accessible supply of thousands of gallons of gasoline. By reason of the gas station's nearness to the streets and more cars and gas stations and, eventually, the pipelines and oil wells, virtually the entire country might explode in a chain reaction. These descriptions of the possible accidents might have led to conjectures that millions might conceivably die in a horrible holocaust.

Near the end of the discussion, the attendees might have comforted themselves with the thought that what they had pictured could never happen, for one simple reason: early accidents of all kinds, from collisions to gasoline explosions, would start taking place at so high a rate that everyone would be frightened off from further expansion. We would surely ban the auto before it could hurt us, they would have concluded.

But we did not ban the automobile, and the great national explosion fortunately never occurred. Moreover, we have accepted the annual injuring or killing of 500,000 people — which deserves the title of holocaust, since the individuals harmed and their families are affected as surely as if the total of annual casualties occurred in a single accident.

Returning from this diversion to the specific dread of nuclear energy, there is another reason to expect that this fear will be with us for a long while. The experts are bound to go on differing in an important area that is apart from their varying value judgments or their individual intuitions about what degree of a gamble is right for the society, considering all the alternatives. We refer to the gray question of how much damage small doses of radiation can do, especially when the amounts are so small that adequate data on the harm are hard to collect and quantify. Individual human body reactions to such small exposures may vary greatly. We are not concerned here with doses whose strength is but a fraction of the exposure we are subjected to with a dental x-ray, or with the incremental differences in radiation each of us receives on the surface of the earth as we travel from sea level to Denver. Rather we are talking about the accidental escape into the air or the ground of contaminated radioactive matter that may be degrading the health of individuals exposed but at such a low level of harm that it may take twenty or thirty years to observe the final effects and prove that health impairment indeed occurred.

For instance, guesses have been made of the amount of radioactive matter that escaped into the atmosphere during the Three Mile Island accident. The number of additional cancer deaths this contamination could conceivably cause was then estimated. In the total area and over the pertinent period of time believed sensible to include, the total of cancer deaths normally to be expected was figured to be around 300,000. The additional cancer deaths the escaped radioactive matter might cause was believed to be about 10. The comparison is then between the number 10 and the number 300,000. Now, perhaps these estimates are off badly. If they are anywhere near right—even, say, by a factor of 10 or 100—this incremental cancer-producing effect is so small and so difficult to measure that we might never know by experience what it actually turned out to be. In the intervening 20-to-30-year period, changes in numerous other factors (for instance, smoking and diet habits, and demographics) will influence the cancer rate in the area far more and will completely mask the effect of the radioactive gases released at Three Mile Island. That does not mean we should ignore small contributions to the undermining of our health. But it does say it is going to be hard to know what we are talking about, and easy for people to argue, if inclined to do so, from varying platforms of individual feelings about risk. Try as we will, in this region of low-probability hazards, experimental and theoretical research will have to be carried on for decades before we can hope to be much clearer about the facts.

Most serious students of this problem express the feeling that dangers of this sort are acceptable, considering everything else about our way of life. When we include in everyday risks those from very low-level pollutants—steady, slight amounts of radiation or exposures to minute quantities of certain possibly dangerous chemicals—the existing number of sources of potential harm, nuclear energy aside, becomes enormous. Fearing and banning all of them would appear absurd to some because the very banning would change our lives and bring in greater health hazards from the basic deprivations we would create. But not all people agree. Some have such high degrees of concern that they would like us to go slow in all applications, or work even harder to eliminate the possibilities of even the most minor of negative effects.

If strong safety concerns about the conventional nuclear reactor are permanent, we can assume that apprehensive feelings will be even greater about breeder reactors. American breeder reactors are still in a developmental stage and many billions of dollars will have to be spent through the 1990s before their economic feasibility will have been established. Today there is less sure knowledge about the environmental safety of breeders than about the present, conventional non-breeder reactors.

The Carter Administration has been trying to persuade other nations to forego the construction of plutonium-burning breeders and to concentrate American research on new types of breeders that will use some other, less dangerous fuel. Evidence of the current confusion in nuclear energy matters is that the President, considered an ally in the camp that regards nuclear energy as a permanent, important component of our total supply, has been adamant in his opposition to the country's chief breeder reactor project (Clinch River). He appears here to agree with those who contend that the project will encourage the proliferation of nuclear weapons because this breeder reactor produces plutonium, the principal critical material of nuclear bombs. Congress has not supported the President's views and has voted to go ahead with funds for the Clinch River program anyway, presumably fearing that not doing so would endanger our future nuclear fuel supply.

The continuing public worries about health and safety guarantee to the government certain continuing roles and missions in nuclear energy, responsibilities that cannot and will not be assigned to the free enterprise sector of the nation. To understand these government functions better, let us be very optimistic about nuclear energy's future for a moment. Let us assume that a very strong majority confidently decides we should go ahead with nuclear energy, believing the need great and the safety and environmental hazards definitely controllable to a satisfactory level. The electric power utilities thus find a favorable climate for speedy approvals of well presented, well designed installations. The bureaucracies concerned work with great speed and efficiency. Court challenges are few and fought off with ease because of the soundness of the approaches as seen by the judges. New nuclear power plants are created at a remarkably fast rate. They soon spew out a generous supply of electrical power at a reasonable price and with no accidents, and more power is promised soon.

This happy development would ease gas and oil requirements and give us more time to utilize coal for synthetic fuel production and to develop solar energy. Even though we would still use a lot of oil and import a good deal of it, our political posture vis-a-vis OPEC and other oil-consuming nations would improve greatly. The threat of lines at the gas station would disappear. Everything generally would confirm that the nation had made a good decision. Under these rather exceptionally pleasant circumstances, what would we see as the right respective roles for free enterprise and the government? For one thing, surely the government would have to continue to set and to enforce safety standards. Our optimism in no way extends to an assumption that the public would be willing to turn over safety to the designers and operators of nuclear plants for voluntary self-regulation on an honor system basis.

If the government were to carry out its regulatory activities in an atmosphere of public confidence that they are being performed competently, it is consistent to assume the government also would be engaged in an exceptional safety research program, tracking down all the data required to understand and assess safety and environmental effects. Otherwise no credible basis would exist for ensuring the hazards are controllable and being controlled. While some studies of these matters might be expected from the technological industries manufacturing or operating the reactors, whatever those industries would do would have to be regarded as insufficient. The government, not the industry, would set the standards and it would have to perform this duty with depth as well as objectivity. It could not be limited to the information turned up by the private sector as it goes about designing for safety in accordance with its different goals and responsibilities. We would strive for excellent cooperation, of course, between government and private sector experts on the technology pertinent to standards and regulation. Working together in areas where doubts exist, they would seek to clarify the importance of any effect upon the public or the environment.

The government would also take the initiative in the matter of waste disposal and transportation of radioactive materials. In theory each electrical generating installation could take care of its own wastes in some manner covered by government regulations. However, the waste disposal problem very likely would be best met by an overall national scheme, from the standpoint of both safety and economics. At a very big nuclear installation located in an isolated area, everything related to generation of power and to the handling, reprocessing, enrichment, and permanent storage of radioactive materials conceivably could take place. Possibly, all the necessary steps could be done well by the private group carrying on that project; however, government standards would still have to be fulfilled under the government's critical, watchful eye.*

---

*The needed permanence of the government's role in waste disposal is demonstrated by another, quite different nuclear waste disposal problem, one exceeding that of

In this regard, consider a project that has been suggested from time to time: a nuclear power generating installation on the St. Lawrence Waterway, at the location of the locks.** The water discharged (necessarily warm) from the nuclear plant would make it possible to operate the locks throughout the year, even during the most severe winters. This would make Chicago and other Great Lakes cities international all-year ports and could yield great additional returns on the investment to the United States and Canada. The proposed location makes it feasible to create a huge installation to provide electric power for a considerable portion of the Northeast while carrying out all waste storage and processing functions with safety and confining all radioactive materials handling to that site permanently.

The expansion of nuclear power envisaged in our optimistic scenario would require that the breeder reactor be available by, say, the year 2000. It is inconsistent to assume large private-capital investment in light water reactors in the face of a possible shortage of economically available uranium in 30 or 40 years. However, as indicated earlier, America's breeder reactor program, which has already involved well over $10 billion, will need more funding of greater magnitude before all of the detailed engineering will be complete and full-scale working units are available. A very long development period, huge financial investment, and government control of health, safety, and environmental standards in a political environment — these form a combination not well suited to the reward–risk requirements of private industry. Conceivably, a change in the nation's attitude toward antitrust actions could allow large corporations to pool funds, thus making available the capital needed for the breeder part of this optimistic nuclear energy program, without government assistance. However, in putting forth this imaginary scenario, we are not proposing to change the entire makeup of America, but only to assume favor-

waste generated by the conventional nuclear fission reactors and yet, strangely, rarely if ever mentioned. The problem is that of storing (and moving) nuclear bombs. Longtime neglect of this storage problem is not to be expected since those most apprehensive about dangers from nuclear reactors are also most anxious, not surprisingly, that the nations of the world ban nuclear weapons. But if we and the Soviet Union and all other nations with nuclear bombs were to agree to destroy them, how would that be accomplished? The most sensible way would be to convert the fissionable material into fuel for nuclear reactors and use it up. But we could not do this if we had already banned and destroyed nuclear reactors.

**The author first heard this suggestion in a conversation with Dr. Arthur Beuche of the General Electric Company.

able developments especially peculiar to the nuclear energy situation we face. Thus, it is sensible to surmise that the government alone, and not the private sector, will continue to be the source for the R&D funds for the breeder reactor. In fact, only if the government pays for the breeders' development and ensures the availability of their enriched fuel output will adequate private capital probably be available for the conventional nuclear reactor portions of the program.

This optimistic story we have just described can help us realize that if nuclear energy is to move forward the participation of both the government and the private sector is needed. If we wish to invent ways to act soundly on nuclear energy, we need to assign proper roles to government and the private sector. We need to obtain the right degree of contesting and cooperation between these two categories of our operating society. To see how to do this, it may prove of value to return first to the other extreme, in which we assume the United States abolishes the nuclear energy option entirely. We stop all new installations and phase out existing installations, accepting the financial losses. If ten years from now we decide we were wrong, we will have lost more than the funds previously invested and the time spent because we will not be able to maintain our competence in the interim, our readiness to move ahead if we so choose. However, in electing this alternate route, we imagine we are willing to accept the risk of being wrong. If it turns out we were right, we will have eliminated the harm to the United States from all of nuclear energy's hazards. We also will have gained the availability for other purposes of the technological resources otherwise devoted to the nuclear route.

Of course, we will never know later what disasters might have taken place, all of which we precluded. However, we might regret not having included the nuclear option, if, as many consider extremely likely, we find the nation greatly handicapped by lack of adequate energy. Missing the nuclear energy output, we might have to accept a slowed economy. Unemployment might be higher and the standard of living lower. As these conditions develop, we might be driven, by the political unacceptance of them, to drastic and unwise action to accelerate other alternatives, spending more of our resources than we should in trying to make other avenues work out. For instance we might mount a crash program to use coal and be led to more deaths — in mining and transportation accidents and because of contaminants in the environment — than would have happened with the nuclear approach. All of these things might happen, and yet we would be hardpressed to decide whether we had been right or wrong — except for one thing. While we might ban nuclear energy in the United States, this would have only a modest influence on what happens in the rest of the world. Other nations are going to go on with nuclear energy until and unless, of course, one of them should experience a catastrophic accident. Then all will be convinced that we made the right decision.

For France, Germany, Japan, Brazil, Taiwan, and others, nuclear energy has a different priority than in America. For these countries the alternatives of petroleum and coal virtually do not exist. The nuclear option is being judged by them on a different scale of balance between the risk of dangerous accidents on the one hand and the risk of descending to a lower level of economic activity and standard of living on the other. In their situations an alternate to a fossil fuel energy supply is imperative. They already are practicing the conservation we are now mainly merely talking about. They drive small cars and have good public transportation. They employ less air-conditioning in summer and lower temperatures in their homes in winter than we. As to risk taking, they recognize their populations have already accepted a heavy toll in contributions to the lowering of longevity from automobiles, airplanes, industrial activity, smoking, and alcohol, and do not plan to remove these risks. They believe disease and hunger need to be fought and the fight is easier when the society is prosperous. Prosperity requires energy. In those countries, the climate of public opinion is more favorably attuned to the trained specialists' assessment of the dangers of nuclear energy. It will be almost inevitable that the majority of their citizens (and more especially the leadership, for not all of these nations are constitutional democracies) will perceive the risks of well designed nuclear reactors as containable and actually reducible to negligible through careful, competent practice.

The nations mentioned are all going ahead rapidly on nuclear energy. France and Germany are developing reprocessing and enrichment technology. France has actually begun assembly of the world's first commercial fast breeder reactor, and by 1985 more than half of France's electric power will be nuclear sourced. Germany, Italy, and Spain reaffirmed their nuclear energy plans after Three Mile Island, or actually accelerated them. The United Kingdom also has strong nuclear technology capabilities. Britain's David Howell, Secretary of State for Energy, has announced that the new Conservative government plans to expand the present nuclear program, citing the Three Mile Island accident results as reassuring: "It showed that when some stupid errors were made and the system was put under great stress, safety was still maintained."[*]

*Science, **206**, October 19, 1979.

A recent conference including 66 non-communist nations wholeheartedly supported the rapid development of breeder reactors around the world. Taiwan will soon be all-nuclear in electric power generation and Brazil is moving forward. Japan is experimenting with an advanced design of a breeder reactor. It has recently made a decision to develop its own light water reactors rather than make additional purchases from overseas. We spent 1979 wondering what to do. In that year, Japan put six new nuclear reactors in service, bringing its total to 21, and announced it will double its nuclear capacity by 1985, with another doubling expected in the early 1990s. While we are hesitating on reprocessing and enriching facilities, Japan has developed its own uranium enrichment plant. More than two-thirds of the R&D budgets in energy R&D in Britain, Canada, Germany, and Japan are devoted to the nuclear area.

France is scheduled to become the second largest producer of nuclear electric power in a few years when its Super-Phenix breeder reactor goes into operation. The French have scheduled starting up new nuclear plants at the rate of six a year for the next five years. By the middle 1980s, France will have 40 reactors and will have cut its oil import costs by 30 percent. Construction of two more breeders is planned to start by 1985. The French also operate fuel reprocessing plants and have developed their own long-term storage methods, which they obviously believe are safe. Giscard has claimed that with their planned breeders used to extend their uranium supplies, France's "potential energy reserves are comparable to those of Saudi Arabia."* The West Germans are considering a nuclear energy acceleration because of fears that France's industries, backed by cheap nuclear energy, will win competitions against German industry, which is still dependent on increasingly higher priced OPEC oil.

Although its program is falling short of announced goals, the USSR started production of electricity with nuclear reactors in 1970 and about 5 percent of its electricity now comes from the atom. The Soviet Union today has over 25 nuclear plants, second only to the United States. A mammoth equipment building program has been mounted for the 1980s and five new reactors are scheduled for 1980 alone. The bottleneck to Soviet nuclear energy plant expansion is not citizen opposition but equipment building and personnel training. Highest priority has been given to the nuclear approach and the 1980 rate of production of new reactors is expected to double by 1985. The Russians have announced a reactor export program, with commitments to Cuba and Libya. The East European Soviet allies are espe-

*Time, February 18, 1980.

cially counting on Soviet reactors. Petroleum deposits near the USSR's western industry are badly depleted. Their known untapped reserves are in Siberia, thousands of miles from their industry. The USSR use of reactors includes generation of steam to heat their cities as well as to produce electricity, which means they regard the chance of a serious accident as so remote they are comfortable with plant locations near dense populations. A breeder reactor is nearing completion and a prototype of a small mobile nuclear reactor steam generator for isolated Arctic cities has been developed and is under test.

If these countries approach the nuclear option with dispatch and resolve, their expertise in that technology will advance rapidly. Even if America does not decide to abandon nuclear energy totally, a mere continuation of our stalemate will cause our technology position to deteriorate via-a-vis other nations. Our know-how gradually will get transferred to other countries as American nuclear experts, lacking work here, will take on design assignments on nuclear projects elsewhere. Companies in the United States with skills and facilities capable of making a contribution in nuclear hardware will accept contracts to provide components for assemblies made in other countries. In 10 years' time, the American contribution will diminish as the technology of the other nations moves ahead of ours. After two or three more decades of importing petroleum we may then shift to buying nuclear reactors from Europe and Japan.

We have presented two extreme scenarios for nuclear energy for the United States. The most likely future, if we don't considerably alter our present pattern, is something in between the two descriptions. We will be into nuclear energy and yet not into it. We will move along slowly with intermittent stops and starts, as exemplified by what is happening on the American breeder reactor project. We have not decided to stop nuclear power but we have made it almost impossible to go ahead with it. In his 1980 State of the Union message, President Carter stated that nuclear power is a required option to oil and gas, but this pronouncement does not produce electricity, and it won't in the face of a hiatus in the process of decision making and implementation.

We have made studies, but we still do not have a solid program for disposal of nuclear waste. The President has proposed a special expediting board to eliminate red tape and try to get decisions more rapidly on all energy issues where approvals by state and federal government agencies have been very slow, but it is not clear that Congress will go along with this. Even if Congress approves, it is not clear that such a board would make very much difference. Its expediting endeavors might simply speed up turn-downs.

If nuclear installations move ahead in the rest of the world and go well there, we eventually will begin to move more rapidly in the United States. We will be bound to accumulate information from the experiences of other nations and this should be conducive to enhancing safety in the later detailed designs of our own nuclear installations. We will come closer to a universal acceptance that nuclear energy is adequately safe if the experience in the rest of the globe seems to justify that confidence. Propagandists against nuclear energy who are now influential because the public is mixed up and apprehensive will use up their value in time as mere propagandists. The social–political environment may change to one in which the American public at large will come to understand that both options — either moving forward on atomic power or banning it — involve risks.

But we can do better right now. The nuclear decision is only one of many we have to learn how to make in the relating of technological development to the society's needs. Let us discuss some possibly practical means for resolving the nuclear question early rather than late.

President Carter has suggested that the energy issue for the nation be considered by the public as the moral equivalent of war. The President's words have not had much effect, since the issue is not a matter of morals at all. However, as in war, we have to make serious decisions more expeditiously. We are a democracy but every national decision cannot be made with the concurrence of all the voters. We have to settle for acting on a majority desire, with proper protection for minority views. We have to have a way to bring all aspects of an argument out before the public in an orderly way, providing proper arenas for full disclosure and discussion. Then we need a mechanism for getting the decisions made and implementing them. On minor items, we can afford to let matters take their course. When issues become important enough to affect the nation's economy and security, we have to find a way to speed up the process.

Let us encourage the private sector of the nation, in which electric power is generated for consumers, to propose nuclear power installations whenever they are deemed the best alternative for meeting the requirements of their customers. If they believe they can meet reasonable conditions of price, safety, and environmental controls, the private suppliers should describe what they want to do and include a detailed recital of how they propose to limit the negatives.

Next, let us arrange that the federal government place in one investigatory unit the responsibility for careful examination of the safety and environmental dangers associated with the entire field of nuclear energy, and with each proposed installation — not approval-granting or decision-making responsibility, just a critique of the proposal. This unit would be a department of the broad investigatory agency on safety, health, and environmental protection (SHEP) we sketched in the previous chapter. In preparing to do its job well, SHEP's nuclear division would sponsor research on health, safety, and environmental aspects of reactors, seeking to apprise itself of the facts and to become as competent as possible in carrying on the examinations totally constituting its mission. The power-generating private sector group, meanwhile, would also carry on suitable R&D as would the various technological companies supplying apparatus. In order to be able to design, and defend their designs from the standpoint of safety and environment, they would make experimental and theoretical studies of their own. The government's and the private sector's derived information would all be made public.

It is important to recognize what the suggested SHEP division would *not* be. It would purposely be given a narrow mission. It would not be asked to decide whether the proposed facility should go ahead or an existing one be stopped. It would concern itself only with detriments and be a source of expertise on them. It would exist to provide protection of the citizens by opposing any activity it believed to be harmful to them.

The problems of decision are those of weighing the advantages, disadvantages, and costs of a proposal against the loss of benefits and the disadvantages of its not going ahead. Often such a decision will hinge on gray areas in which adequate data are not available, where gambles must be taken, where judgments must be made even if based on an inadequate amount of information. The investigating, negating SHEP agency would not be set up to make these tradeoff analyses, to compare the alternatives of approving or not approving nuclear energy. It would not be expected to be expert on the consequences to the United States of going to other energy alternatives or doing with less energy. It would not be in the business of deciding how severe for the society an energy shortage will be. It would not concern itself with the possibility of wars over world petroleum resources. It would not be involved with the question of what would happen on the unemployment front, to the poor, to the coal miners, or to the lambs in Wyoming if nuclear energy were not utilized. The proposed investigatory agency would be involved only with the existing and potential hazards of nuclear energy and the applying of its expertise in a confrontation with any group that wants to go ahead with a nuclear installation.

As to existing nuclear operations, this agency would be charged with inspection. It would recommend standards that it would use as guides and alter them with time, as it saw fit. If it were to arrive at a conclusion that a particular nuclear operation should be changed or halted in the interests of safety or environmental protection, it would so propose. If those running the given operation were to agree and act on the recommendation (perhaps because they decide it is deserving, or maybe because they think they could not win before objective judges), no problem of decision making would exist. If not, a decision would have to come from somewhere. As we explained in the previous chapter, we are completely willing — in fact, desirous — that these two groups, one private* and the other governmental, with two substantially different kinds of objectives and missions, be in an adversary position. To oversimplify a bit, we have imagined a protagonist and an antagonist for the proposed installation. We want them to confront each other.

Having now imagined two basic sides to an important issue, let us set forth a way in which the two sides will be caused to put their cases forward completely and fairly so that an objective decision can be made. We shall call on a third party, an impartial judge, a decision-making entity such as we described more generally in the previous chapter. Congress should pass legislation creating a decision board for nuclear energy alone, its members appointed by the President and approved by the Senate. They would serve for a substantial period, say, five years, and, like members of the Supreme Court, would be impeachable for cause but could not be fired by the President. They would be full-time appointees, devoted entirely to serving on this board. When a nuclear issue is raised, the board would listen. It would hear out those who propose to engage in nuclear activities and those in the government's hazards investigating agency. It would also hear from various representatives of the federal government, states, counties, and private citizens, and order up tradeoff analyses when they are deemed desirable. After a reasonable time spent in hearing arguments from all parties, with an adequate amount of interrogation and deliberation, the board would make a decision to let the project proceed or not. Its use of the confrontation between the project's advocates and SHEP, the government's hazards investigation agency, would be analogous to the confrontation our courts of law allow between a government prosecuting attorney and a defense attorney.

*Of course, nuclear reactors for generating electricity might be operated not only by the private sector but by a city- or county-owned utility or the federal government (for example, the TVA). This would not alter the role of the purposed decision board.

The judge in a court case is supplied by the government to make decisions over the government prosecutor and the private defense attorney. Similarly, the proposed nuclear decision-making board would make decisions over the government's safety and environmental hazards agency and the group seeking approval.

Just as the judge in a court case engages only in judging and decision making, the nuclear decision board would do nothing else in the nuclear field except decide the issue before it: should an installation or operation be allowed to take place? We seek, from the government and the private sector, that they cooperate in a decision-making procedure. While on opposite sides, like a defense attorney and a prosecuting attorney, they are to act within a cooperative framework in which they accept the court of law, its procedures, and its decisions.

It may very well be that the decision board on occasion would say it does not think the data needed for decision have been adequately presented. There can be many reasons why the board might turn down a proposal and suggest to the proposers that they come back again. We want the board to reject the proposal promptly if it is going to do so. Conversely, if it says to go ahead, we do not want that decision to be disturbed by those who feel differently. In this sense, we want the board to act like a supreme court. Anyone who doesn't like a decision of the Supreme Court can try to get the Constitution changed. Or a way can be sought to come back to the Supreme Court with a different twist in the case already presented. Meanwhile, life goes on based upon the Supreme Court decision.

The legislation creating the nuclear decision board would have to spell out that its decision-making ability overrides previous legislation dealing with nuclear matters. Since the board would be appointed by the President and confirmed by the Senate, the political aspects of nuclear energy and the differences in value judgments of various constituencies would all manifest themselves, first in the election of the President, and then in the pressures put on him by various constituencies to influence his choice of individuals to nominate for the decision board. The Senate also would reflect the country's varying opinions about nuclear energy and other considerations, so the senators' examining and accepting of nominees would mirror these public feelings. The terms of the initial appointees would be staggered. With five members serving five years each, an opening would occur once a year or oftener on the average, and this would alter the composition of the board and tend to keep it reasonably current politically but not subservient to rapidly changing political fads.

It is important to distinguish between the organizational, decision-making scheme we have just described and two other organizational entities with which it might be confused: (1) the present Nuclear Regulatory Commission, and (2) the President's proposed Energy Mobilization Board. The first is an impractical and unworkable combination in one agency of the two units we have called for: an investigatory agency and a decision board. The fundamental flaw of the present NRC is that it is both the judge and the prosecutor. In effect, it receives a proposal for a nuclear power application, it provides the case against it, then it decides between the two sides, one of which is its own. It is expected to watch out for the negatives of nuclear reactors, to guard the public against unsafe operations, but is not charged to make tradeoff studies, to compare the negatives of nuclear power against the negatives of no nuclear power. Yet, in some ways, it also conceives itself of being pro-nuclear, a party to bringing nuclear energy to the service of the American people. When it does this, it has a serious built-in conflict of interest.

The President's proposed Energy Mobilization Board is an "expediting" board, one intended to accelerate energy installations by speeding up bureaucratic actions now causing delays, perhaps bypassing some existing safety and environmental legislation. Unlike the decision board we have here proposed, one charged to make tradeoff analyses and objective decisions for the broad best interests of the nation after judging the claims of opposing advocates, the President's suggested board would exist solely to push for quick action. It is proposed on the premise that we want to expand energy facilities and it exists to help bring that expansion about — doing it, if necessary, by trampling a bit on those who oppose the expansion.

Returning to the new decision board we are advocating, would it be constitutional? The board could not be given the powers of the courts, the Congress, or the executive branch as it applies to nuclear power. The board could not cause an installation to take place which deprives any state or citizen of constitutional rights. It could not even say whether nuclear power should exist and if so at what level of contribution to our total energy requirement. It would merely sit in judgment on the creation of new nuclear power generation installations or the altering or stopping of existing ones. Hence, legal paths would remain available to those who oppose an installation. They could seek action in court to prevent it even if the board has allowed it, or to obtain the right to proceed even when the board has rejected it. To be sure, some of the board's actions might be brought to the courts, even the Supreme Court, for review. Such court actions would help define the powers of the board but, with proper legislation creating it, the board would not be thrown out. The board, if created, probably would be able to speed up the timetable and improve the quality of decision making on nuclear power generation in the United States while also providing an objective arena for efficiently pulling together all sides of the argument.

Many decisions about nuclear installations, continuing or new, would not have to go before the decision board. The operators or would-be operators doubtless would prefer to settle "out of court" with the investigatory agency. If the agency demanded safety improvements, the operators' group might decide to incorporate them rather than take the issue to mediation. The two confrontation groups would be in constant communication trying to resolve their differences, minimizing those that would require going before the independent third party, the decision board.

On the assumption that both SHEP and the decision board were to be caused to exist, let us summarize appropriate roles for government and the private sector in nuclear power generation. The private sector (or a government-owned utility in the electric power business) would go about making its proposals to create nuclear power installations and would operate nuclear facilities once built. The nuclear industry would be free, of course, to carry on such private R&D as it may choose, in order to improve its capability and competitive position. It would sponsor research in safety and in environmental protection to aid it in the design of superior installations. The government, through its nuclear hazards investigatory agency, SHEP, would take the initiative in providing protection of the nation against nuclear dangers. SHEP would engage in contraactual activities with centers of know-how (the industry, universities, non-profit and its own government laboratories), to investigate and understand those facets of nuclear technology having to do with safety and environmental effects. Whenever any differences surface, the decision board would make decisions, always seeking to act in the overall national interest. The sponsoring of further advanced R&D on breeders or nuclear fusion or other nuclear phenomena would continue under the Department of Energy or the National Science Foundation or the Department of Health and Welfare or other agencies now engaged in such efforts. No organizational changes would be needed, at least not any resulting from the creation of the proposed SHEP and decision board.

8

the synthetic synthetic fuel controversy

The United States imports billions of barrels of petroleum each year, exiling several tens of billions of good dollars from the country to pay for them, while sitting on deposits of coal whose stored energy content substantially exceeds that of all the known oil reserves of the Near East. Using existing and extendable technology and the cooperative effort of government and private industry, we could arrange a bounteous supply of liquid and gaseous fuels for centuries from these deposits.* The resulting energy would cost more than today's petroleum-based equivalents or that obtained from the straightforward burning of coal, but surely foreign petroleum will rise in price steadily to equal the cost of the cheapest plentiful substitute. With adequate energy, even at a higher price level, in contrast to meager rations at an artificial, subsidized low price, America could prosper, according to energy economists in government, industry, and academia who have scrutinized this tradeoff. The benefits of growth, flexibility, and stability of an economy based on plentiful energy would outweigh the penalties of higher cost. The added funds expended would remain in the United States and our dangerous dependence on foreign sources would be reduced.

However, synthetic fuel from coal is a controversial issue. One aspect of the controversy—technological feasibility—is synthesized

---

*When fossils in liquid form (petroleum) are taken from the earth and, using complex operations, gasoline is extracted from them, this fuel is labeled "natural." On the other hand, if fossil solid matter (coal) is the starting point, then the identical gasoline, molecule for molecule, which the appropriate process turns out is called "synthetic." The government also uses the name "synthetic" for fuel from biomass or oil shale.

from ignorance and is dispelled by noting the progress in South Africa, where synthetic fuel has been made for a quarter of a century. During that time the South Africans have refined the process, increased efficiency, and learned a great deal about pollutant control. Their process is conventional, an improvement over the technology the Germans used during World War II to provide fuel for tanks, trucks, and airplanes. The methodology was advanced by South Africa with the aid of American companies. Heated coal, steam, and oxygen are combined to yield a gaseous mixture of carbon monoxide, hydrogen, and methane. These gases, under compression and at high temperature, are passed over an iron-based catalyst to produce oil and fuel gas. The oil is then refined into gasoline. A ton of coal makes between one and two barrels of oil.

The South Africans are already consuming some five million tons of coal a year and are approaching a production of gasoline equal to about 10 percent of their needs. They expect to increase that figure to 30 percent in less than three more years and then go on to higher levels. The costs are said to be only modestly above OPEC prices. They have invested about $10 billion so far. Their environmental standards for air pollution are probably less severe than those we would apply to American synthetic fuel plants, but there is little to suggest the ultimate has been reached in their process from the standpoint of efficiency or cost or environmental controls. They have designs, it is said, that will almost eliminate liquid waste.

This process, applicable to mined coal, is not the only possibility for obtaining liquid and gaseous fuels from coal. Another approach is to extract these products from coal in situ (while still in the ground). Coal gasification tests have been conducted successfully by penetrating the underground coal seams to produce cavities and controlled fires within them. The resulting hydrocarbon gases and fluids are piped up and refined into high-BTU pipeline gas or jet fuel, diesel fuel, or gasoline. Other conversion processes based on coal do not have fuels as their output, but rather commercial chemicals now being made from petroleum. For example, Eastman Kodak is building a plant to produce acetic anhydride, which is used in manufacturing photographic film, fibers, and plastics.

With the abundance of coal in the United States and the realizable technology, it would seem foolish to overlook liquid and gaseous fuel from coal* for development as a substitute for petroleum imports.

---

*We have featured coal as the example in this chapter. For most points made we could just as well have used oil shale.

However, America does not have such a synthetic fuel program, at least not in a serious sense. We have not even reached the definite conclusion that this route, considering its negatives and the availability of other alternatives, deserves extraordinary attention. We ought to be able to decide what to do about this energy option and then go do it. Instead, the issue of the establishment of a United States synthetic fuel industry based on coal remains confused and unsettled. It is an excellent example of the nation's ineptitude in arriving at decisions to balance environmental and safety considerations against critical supply needs and to define proper roles for government and the private sector.

The manufacture of synthetic fuel (syn-fuel) from coal illustrates these shortcomings even better than nuclear energy. The latter, as we observed in the previous chapter, spectacularly reveals the ineffectiveness of U.S. decision making. However, as an example of the general problem of technology–society interfaces, it suffers from domination by only one factor, the fear of cataclysmic accidents and massive harm to the population. Nuclear energy, as an area for decision and control, is stalemated because of the strength of this special fear. Of course, coal-based synthetic fuel installations and their supporting infrastructures of mining and transport are replete with frightening possibilities for environmental degradations, health penalties, and high accident rates. Some nuclear advocates and other, more independent analysts assess these hazards of synthetic fuel from coal as greatly exceeding the real nuclear dangers. However, the nation's voters were not born with nuclear reactors operating around them. When they were children they did not observe their parents calmly accepting nuclear risks. The dangers of coal have been with us as long as we can all remember. Coal soot and coal mine deaths are old and familiar evils, so we are able to deal objectively with coal. We recognize its negatives but, regardless of how serious they may be, we tend to regard them as analyzable and measurable, and the cost of adequate controls as realistically estimable. Maybe coming generations will be better able to handle decisions about nuclear energy, but for now, coal is different from nuclear energy as a decision area.

Moreover, nuclear energy is of limited value for illustrating the way government and the private sector should relate so as to put technology to work wisely. A totally private-sector or risk-capital route to nuclear energy has been and will remain a closed road. In contrast, until very recently coal has been a private-sector domain with little government initiative to form projects and sponsor R&D.

With the fright and frenzy factors less fatal and a free enterprise future not forbidden by the force of past fashions, the synthetic fuel alternative should offer us an opportunity to focus on some key dimensions of the decision-making and organizational problem. We should be able to ponder intelligently such reasonable, unemotional questions as: Does the nation really need a big program to produce synthetic fuel from coal? What is the tradeoff of need versus potential harm? Should we rely on the free enterprise system to generate an appropriate program, large or small? Should the government launch a crash program to develop high-volume production of synthetic fuel from coal at the earliest date regardless of what the private sector does? What is the best role for government? Why must it be so hard to decide these questions? Why must we synthesize a controversy about it?

If a synthetic fuel option based on coal makes no sense either technologically or economically, and is inferior to other practical available alternatives, the private sector certainly won't invest in synthetic fuel. If political requirements nevertheless compel us to travel this route anyway, the government will have to drive the bus. On the other hand, if the option is attractive to the private sector, presumably because market supply and demand forces favor potentially good returns on investment, it will develop through free enterprise without government aid—unless the government intervenes, purposely or inadvertently, to prevent the market from forming naturally. Whether a market for privately financed synthetic fuel occurs, or a government entry takes place instead, or both, will depend on the working out of a bundle of economic, social, political, and technological factors. Included among these, perhaps in the prime position, is the matter of competitive alternatives. Syn-fuel (from coal) may be the way to go, to some appropriate extent, for some energy requirements in some geographic areas. For other applications in other places, superior options may exist either for private-risk investment or for a government-sponsored effort.

Competitive approaches to meeting our energy needs are not lacking. All have their positives and negatives, their strong advocates and detractors. For instance, it is claimed by many petroleum professionals that installing submergible pumps and carrying out so-called secondary and tertiary recovery operations on existing domestic oil wells would yield more oil than the equivalent of a decade or more of the largest conceivable scale of syn-fuel activity. Furthermore, it is maintained, this could be done at far lower costs, with greater certainty of success and no environmental harm. Finally, the advocates of more attention to this option believe it can best be caused to happen without government expenditure or involvement. All the government has to do is lift all price controls, stop allocations, and abandon excess profit tax ideas, thus allowing the greatest of motivations for investment in the active free market that will result.

None of this involves newly found petroleum. Some sophisticated analysts of the problem tell us that, if the market were set free, exploration on American-controlled land and offshore areas would increase and in 10 years would locate more added supply than syn-fuel plants from coal could possibly provide during that period, again with less investment and environmental impact. A free market would also increase investment in searches for petroleum elsewhere in the world. Those optimistic about this potential point out that only a tiny fraction of the earth has been examined and that greatly increased world oil discoveries, even if not under American control, would vastly improve our energy situation. It is hard to dismiss the serious professionals who are optimistic about our extracting much more petroleum from the earth if all-out efforts at discovery were encouraged, when the facts they cite are considered. As an example, according to estimates by the U.S. Geological Survey, 30 billion barrels of undiscovered oil exist in Alaska and 75 billion cubic feet of undiscovered natural gas remain untapped in that region. In the past century some 3 million oil wells were drilled in the entire world; three-quarters of these were in our 48 contiguous states. By comparison, the rest of the world lies largely unexplored. Even in the United States, vast areas (over 500 million acres) owned by the federal government are not being tested. Present leases granted by the government for offshore exploration cover less than 3 percent of the entire area. In size (and, some geologists say, in oil-find promise) it is equal to our land territory. The area so far drilled has produced almost 10 billion barrels of oil and 50 trillion cubic feet of gas.

Gaseous syn-fuel from coal was estimated at the start of 1980 as costing in a range more than twice today's natural gas prices. Those who search for natural gas deposits argue that if they were permitted this double price they could easily outproduce the syn-fuel plants. Higher prices make it economically feasible to consider new areas and to drill deeper. For example, the west has a strip running from the Mexican border through Utah to Canada with complexly formed, densely packed terrain. The gas deposits here are very deep, requiring the latest technology to explore. This so-called unconventional gas is found also in the Appalachia shale country, in underground caverns along the coastlines of Louisiana and Texas, and in many coal seams in various parts of the country. There seems little doubt that large amounts of gas are there but it may be costly, and it is not clear how costly, to get the gas up. The probable reserves contained in all these added potential sources have been estimated at over 100 trillion cubic feet, a figure comparable with presently identified reserves of more conventional, readily recoverable natural gas deposits.

Nuclear power, despite its enemies, also has supporters who claim for it a combination of economy, cleanliness, and safety greatly superior to syn-fuel. Others push solar energy in all its forms, including direct conversion of sunlight to electric power through solar cell panels and the production of fuels from biomass. Some of those most interested in increasing reliance on coal believe the syn-fuel approach, turning coal into liquids or gases, is not as sensible as the conventional burning of coal with improved control of the process by advanced technology, to enhance efficiency and limit environmental degradations.

Even larger than our coal deposits is the huge store of oil—well over a 100-year supply—locked in the rocks of the western states, mainly Colorado, Utah, and Wyoming. About 1½ tons of shale rock must be mined and heated to produce a barrel of oil. The process is more straightforward than obtaining oil from coal, but the conventional approach requires huge amounts of water, scarce in the west, and enormous piles of ash-like rubble are left to be dealt with. To get a million barrels of oil, several times that much water is used, and a million tons of solid waste result. To avoid mining and the problems it brings, new approaches are being tried to create underground cavities within the shale. Controlled fires set in the cavities would cause a separation of the oil from the rock into pools from which the oil could then be pumped to the surface. If this scheme can be made to work, oil from shale may rank well above oil from coal. Some professional proponents believe from initial experiments that success is highly probable. Others are concerned that undesirable salts, perhaps even arsenic, may be released into the ground water of the shale regions by this *in situ* process.

The biomass alternative cannot readily be ignored by those who press for a giant program of syn-fuel from coal because biomass has so many dimensions and sub-alternatives, some of which we discussed in Chapter 2. Biomass can be burned, fermented, or reacted chemically with other materials to release energy. For instance, quite apart from the plants that produce hydrocarbon fuel molecules directly, in American forests the trees that rot away each year have an energy potential nearly double that of all the nation's imported oil.

Sugarcane residues could be burned to generate electricity for Hawaii. Farming kelp in the Pacific Ocean could lead to production of methane gas in very substantial amounts. Burning peach pits in California could make enough steam to handle the canning of peaches. Peanut shells, spoiled wheat, corn husks, cattails, and cow dung are all candidates for biomass energy generation. Letting the sun shine on water, if it contains the proper chemical catalyst, will produce hydrogen and hydrogen peroxide, both useful fuels. There can be little doubt about basic technological foundations for fuel from biomass because making alcohol, one useful fuel, by the fermenting of agricultural crops is as old as agriculture. Furthermore, efficiency can be essentially as high in small installations as in very large ones. Almost any size farm, if properly equipped, could produce economic fuel this way, assuming present fuel price trends continue.

The burning of garbage to produce energy is taken very seriously by some and it is not a total waste to consider some of their reasonings. In backward societies, animal wastes have been used for fuel for centuries. Even in this country, before the energy crisis surfaced, 20 garbage-to-energy plants were already operating, the heat of the burning garbage producing steam to drive conventional electric power generators. The nation has many millions of dairy cows. Using available technology on the manure these cows produce daily, theoretically we could create the equivalent in methane gas of a million gallons of fuel oil. In addition, the leftover waste could be used as fertilizer. The price of other techniques for producing energy used to be much lower than that of obtaining it from garbage. However, recently the mayor of New York announced a deal with a private company to construct a $100 million garbage-to-energy complex. New York City accumulates more than 22,000 tons of waste every day, and if nothing else, new federal anti-pollution guidelines have presented New York City with a crisis as to what to do with it. The mayor may have the answer.

Of course, there are detractors of biomass who hint its enthusiasts have not made the right calculations. If corn is grown and turned into alcohol as a contributor to "gasohol" for automobile fuel, then, depending on how it is done, the efficiency of the process might be so poor that more energy could be consumed overall in planting, watering, harvesting, and processing than the amount that ultimately would reach the user. If large-scale acreage is removed from other possible applications to provide energy, this is an environmental impairment. Even if the proper plants are used to create large-scale energy in the biomass approach, water availability may be a limiting factor. Also, the soil may lose its nutrients after a number of crop seasons. If fertilizers and pesticides made from fossil fuels are then applied, this is cheating a little because raising crops to produce energy, as they become heavily dependent on added chemicals, may result in a net loss of energy. In this regard, we note that some underdeveloped African countries that use solar energy in the form of a wood crop are now experiencing an energy shortage more drastic than ours because their forests are becoming deserts.

But there are even more energy alternatives. The total power potential from undeveloped small and large dam sites is estimated by some experts at 200,000 megawatts, which is about 40 percent of the nation's electricity supply. It is claimed by still others that, by the year 2000, geothermal energy, the heat stored in the earth, could provide another 5 percent of the nation's needs. There have been estimates by engineers that enough practical windmills could be erected by that same year to yield still another 5 percent.

Finally, we should not overlook a suggestion sometimes made* which recommends that the government buy small, high-MPG replacement cars and give them free or at a reduced price to the present owners of old, large cars or gas-guzzlers, turning the latter into scrap. This would remove about 50 million energy-wasting cars, largely owned by the lower-income half of the nation. The cost would be around $100 billion, about the same as the price of creating a major syn-fuel capacity equal to the annual savings in gasoline the proposed auto replacement program would be able to effect. To buy the new cars would be faster and surer and would improve the quality of the air while eliminating the certain environmental degradation a syn-fuel industry this large would produce. Perhaps not to everyone a wholly serious proposal, but one that makes a point.

Along with the foregoing list, and other energy supply alternatives to compare with syn-fuel from coal, we must consider conservation as an answer. With expenditures of funds possibly well below coal syn-fuel capital requirements, we can insulate homes and commercial structures, improve urban transportation and industrial processes, and buy tens of millions of those new, energy-stingy cars we just discussed — all of which would save more fuel than synthetic fuel plants could produce even with the biggest program imaginable. In fact, the conservation concept introduces so many key issues in technology–society interactions that we shall devote the next chapter

*See, for instance, a letter to the editor, New York Times, January 1, 1980, by Amory B. Lovins.

to it. Of course, even conservation measures can create health hazards. It has been claimed by the EPA, for example, that if, in stopping heat leaks, ventilation in all American homes were cut in half, lung cancer deaths from radon gas would rise by 10,000 to 20,000. Radon is produced when the uranium always present in bricks and concrete decays. Similarly, foam insulation for houses contains small amounts of evaporating formaldehyde which has produced tumors in rats. The EPA is not certain how to attack the problem of indoor pollutants based on building materials because, among other things, it is not sure it has the responsibility to set the standards for this category of hazard.

Like synthetic fuel from coal itself, all of the alternatives possess an abundance of negatives and doubts as to success and practicality. To illustrate this, let us focus briefly on the solar energy alternative. The case made for solar energy is usually that it is endless and free. The basic source, the sun, is permanent. Also, solar energy is inherently totally safe and lacking in environmental degradation characteristics. The problem so far has been that, except for a small fraction of applications, it is expensive. The known and developed techniques for changing sunlight directly into electricity for a home, building, or industrial process involves relatively expensive apparatus, particularly if the energy application requires steadiness so that storage means must be available to cover the periods when the sun's rays are not present or are too weak. The economies of home hot water or space heating by direct solar radiation work out better, but only in a small fraction of all American residences and buildings.

Enthusiasts for solar energy advocate the removal of these limitations by an all-out R&D effort. Sometimes a crash scientific effort, or a cost-reduction technological program, will achieve success and sometimes it will not. It cannot be known ahead of time what a huge research effort would accomplish in the solar field. It is speculative whether we will ever learn how to make solar energy economically competitive with alternate approaches. However, hundreds of millions of dollars already have been appropriated and over a billion dollars will be spent over the next few years to investigate various possibilities and to search for practical solar energy methods. Perhaps an exciting scientific discovery or technological breakthrough will come of all this, but none is guaranteed.

Some perceive the solar enthusiasts as universally given to exaggerated claims based on quasi-religious faith rather than to facts founded on scientific, engineering, and economic fundamentals. This by itself is an exaggeration. Most reputable scientists and engineers with involvements in solar energy are quick to recommend that it would be unwise to rule out other energy options on the certainty that practical and high-capacity solar energy will become available. They realize it may be learned that the sun will never work out as a major United States supply source. They would be surprised only if it turned out to be an insignificant option.

At least it seems hard offhand to quarrel with the idea that solar energy is inherently safe. But some do. A few hazard sleuths have noted that solar panels on the roof of a house have to be cleaned regularly to be effective. They postulate some tens of millions of homes in the sun belt equipped with solar panels each to be cleaned several times a year and arrive at a figure of some 50 million householder ladder trips to the roof annually. They then take the odds of falling off the roof and breaking one's neck, or at least a leg, and arrive at a number of deaths and casualties exceeding our annual automobile casualties and approaching the often described, if low-probability, nuclear catastrophe.*

Those interested in discussing the future of technology in relationship to the society often can be identified in one of two groups. One is suspicious of further technological advance, if not actually hostile. This group believes wwe should spend less on technological developments. They see the present technological society as generally unsatisfactory and believe more technology will lead to more disbenefits than benefits. The other group is enthusiastic about technological advance and believes that only through more R&D can we find ways to remove the ills that accompany technological activities. Many of those in this group have the intuitive feel that with ample R&D a solution can be found to almost any problem. Solar energy supporters without technical background are often found among those critical of technological advance, but they exempt solar technology development from the rule that technological advance brings more harm than good. Thus, as long as solar programs are not described as the certain answer to all our energy problems, these programs can count on more generous support and less opposition than is directed at synthetic fuel from coal.

Solar energy is already receiving a substantial amount of subsidization. Two-thirds of our states now have grants or income tax, property tax, sales tax, or loan incentives for anyone desiring to put a solar panel on a roof to connect to a hot water tank. While California is the

*Although these comments go beyond his, I first heard this hazard described by Prof. John Holdren of the University of California, Berkeley, in March 1979.

acknowledged leader in solar energy incentives, its program is quite typical. Like thirty other states, the increase in the value of the building as a result of installation of solar devices is exempt from property taxes. California homeowners may subtract 55 percent of the cost of a solar system from their income tax bills and spread the deductions over several years. In 1977, 13,000 claims were allowed at an average of almost $600 per claim. Homeowners are eligible for federal tax credits of several thousand dollars to help underwrite the cost of installing solar energy systems. The Military Construction Act of 1979 requires that solar heating systems be installed in new military housing and 25 percent of other new military construction wherever feasible.

Proponents of solar energy have argued that this government subsidization is not nearly enough and should be increased at least until solar energy is treated fairly (as compared with nuclear energy, for example, which has had tens of billions of dollars in government financial support). Others, without expressing any negatives toward solar energy, argue that we should stop any form of subsidization of all energy approaches, rather than add more. They think all government sponsorship is unnecessary interference with the superior, efficient allocation system the competitive market would otherwise provide the nation. As they see it, if the energy alternatives compete in the marketplace for investment funds and consumer acceptance, the right investments will be made and the right alternatives chosen for the numerous and varying requirements and conditions. Different energy sources then will be judged best according to economics, application, geographical location, timing, and magnitude of requirement. All the government need do is set reasonable environmental and safety standards and remove all pricing, allocation, and other controls that handicap the free market's workings. The private sector would then take care of our energy problems.

This is a tempting recommendation from some standpoints. If there were a free market, some energy supply approaches might well blossom more rapidly than they are now doing and would surpass certain competitive approaches. The price of energy surely would rise to true world supply/demand market levels. We would all be encouraged to conserve more, it would make sense for industry to alter its processes, and the purchasing of home appliances and automobiles would be based more than now on conserving energy. Shortages and the possibilities of intermittencies and sudden cutoffs in energy supply would be so penalizing both to industry and the consumer that all would favor the development of further alternatives. We would all become more interested in mass public transit. Individual farmers soon would find they could buy apparatus on the market that would help them use their decaying trees, crop mash, and animal manure to create energy. Or they could purchase and install windmills, or convert crops or agricultural waste to alcohol in purchasable stills for their own use or sale to others. Householders would be more aggressive in conserving heat through the installation of insulation and would become more interested in solar equipment additions to their homes.

However, the government shows little sign of a serious interest in leaving energy alternatives to the free market. The government certainly has subsidized the nuclear option and if it adds to this a large-scale crash program on synthetic fuel from coal, as the Carter Administration has been trying for some time to do, these major interferences with the free market could prevent other alternatives from emerging. Thus, if it later turns out that solar energy's proponents are fully justified in their advocacy, the government's pushing hard on syn-fuel from coal would be wrong. On the other hand, even if the solar enthusiasts are madly overoptimistic, it still does not follow that the government would be doing the right thing by subsidizing syn-fuel. In any case, the contest among alternatives is simply not going to be terminated very soon and may go on for decades, especially as the government enters the picture in a big way, spending funds or guaranteeing loans to advance first one option and then another. The more the government takes money out of the private sector to finance government programs, the less that sector can contribute. It then has less financial means to pit expertise and hard economic-technological considerations against the political pressures that are always the driving force of government decisions on the directions to take in energy development.

Few alternatives for imported petroleum have as many negatives citable against them as a large-scale program to produce synthetic fuel from coal. The process of synthesizing the right combinations of the carbon in coal with hydrogen to create gasoline creates a great quantity of other molecules as well. Many of these unwanted, incidental outputs of the process are cancer causing. Coal syn-fuel plants are dirty, require lots of water—hundreds of gallons per ton of coal used—and need to be close to large coal deposits to be economical. They probably cannot be permitted near large population centers or each other, if stringent health and environmental standards are to be maintained. There will be local resistance to coal gasification and liquefaction plants because they pollute the air with fumes from the burning and the traffic to bring in the coal generates dirt and noise. Just how much energy would be expended to build the plants, roads, railroad equipment, tools, and machinery to create the syn-fuels industry is not yet clear. Arriving at a supply level equal to OPEC oil purchases would be a formidable task for a syn-fuel program. It would require over 50 installations, each costing several billions of dollars.

Among the important environmental effects to consider in decid-ing how far to go with synethetic fuels based on coal is the greenhouse effect referred to in the previous chapter. There is fear that steadily increasing carbon dioxide in the atmosphere from burning fossil fuels can create disastrous warming conditions on earth. Carbon dioxide lets the short-wave radiant heat energy through from the high-temperature sun, but it tends to block the longer-wave infrared heat emanating from our lower-temperature earth, thus raising the earth's temperature. Almost everything we do with energy, from burn-ing fossil fuel to building highways and applying fertilizers, increases the carbon dioxide in the atmosphere. A lot more of this gas is released from production and combustion of syn-fuels, unfortunately, than from direct burning of coal. A major syn-fuel program could cause a doubling of the carbon dioxide concentration in the atmo-sphere and this might warm the world's average climate by as much as 2 or 3 degrees Centigrade, with higher increments at the poles. The result might be that the ice caps would melt, increasing ocean levels enough to inundate most coastal cities of the world. Some scientists believe such a doubling might occur during the first half of the next century if we keep increasing our use of fossil fuels at the present rate of 4 to 5 percent per year.

Other scientists contest this postulate and believe an atmosphere rich in carbon dioxide would stimulate more rapid growth of plant life. The increased photosynthesis activity would be nature's way of neu-tralizing excess carbon dioxide. Even if the greenhouse effect does prove to be serious, carbon dioxide emissions could conceivably be cut by treating the coal to remove pollutants before the fuel is burned. Solvent-refined coal results in lower carbon dioxide emission, closer to that resulting from the combustion of oil. The controversy goes on.

Obviously we have many alternate approaches to increasing our energy supply. But none is a sure thing. Each has its pluses and proponents, its negatives and detractors. It would be very unwise to count totally on any one of the options, or on conservation alone. The nation needs to couple conservation with a variety of sources, old and new, that might enhance the supply in different ways for different uses. It would be wrong to launch a super-crash effort of synthetic fuel from coal, even though we have a lot of coal, if in the course of doing so we violate reasonable environmental standards and neglect to provide respectable funding for other alternatives. It would be delinquent to rule out synthetic fuel from coal entirely. Accordingly, let us assume we wish to provide ourselves with the option of coal syn-fuel, to quickly refine our know-how so that we can move fast on a next step of higher production if we then so choose, or can stop without great loss at some low or intermediate level of production if other approaches turn out better. How can we best arrange for a syn-fuel project along these lines? How do we use the free market most effectively? What is the best role for government?

To answer these questions let us first recognize that the government has little competence with which to perform the details of a large-scale technological effort. The government can attract a few outstanding managers, scientists, and engineers for the top jobs whenever it embarks on a mammoth technical project, but building a staff with top-grade broad expertise is another matter. There will be no comparison in competence — for laying out, developing, designing, experimenting, and bringing to success a synthetic fuel project — between those the government can recruit as in-house executives and technical experts and those American technological industry can assemble. The expertise to provide technological innovation is concentrated mainly in the private sector, that is, in technological industry and the universities. This is especially true of the kind of technology involved in the creation of gasoline from coal. It is in the private sector that one finds expert knowledge of basic physics and chemistry and the necessary experience and know-how in applying science and technology to meet specific market requirements. The technological industry has systems for arranging financial backing and seasoned management to carry out successful implementations. The government team, in contrast, would be burdened constantly with politically based directions and encroachments as it tries to proceed. The government is even lacking in the competence we would like it to have when it evaluates and compares syn-fuel from coal against competitive approaches.

The more deeply the government directs the details of the technological development of syn-fuel, and the greater the dollar magnitude of the government program, the higher the danger of a distorting slowdown and impairment of the nation's program to enhance its energy supply. In particular, any proposal for massive expenditures by the government to develop synthetic fuel — $100 billion or more has been suggested for a new government Energy Security Corporation — can only be financed by taking this capital out of the private sector. This amounts to the government's removing financial backing from the industry professionals and putting it simultaneously into the hands of government amateurs. There is probably a place for synthetic fuel in our list of energy sources but precisely what, where, and when should not be determined entirely by a government agency operating through political reactions. Especially questionable is the enlarged energy agency the government would have to build to carry out a crash synthetic fuel program. Already the Department of Energy's $10 billion of annual expenditures are greater than the total profits of the 10 largest oil companies, yet it is proposed that mature organizations of skilled professionals be drained of available funds in order to finance a new government organization led by political appointees.*

An example of the way the government can frustrate progress rather than generate it is exemplified by some suits in which various agencies are fighting each other in federal courts over policies in the use of railroads to transport coal, a key consideration of any synthetic fuel program based on coal. The Administration and its Department of Energy have advocated strongly the substitution of coal for oil in power utilities. In Texas, in response to this pressure, many utilities have switched entirely to coal-fired power generation. City Public Service in San Antonio started building its new $250-million generating plant in 1973 with the understanding that the railroads would deliver Wyoming coal to the boilers at under $8 per ton. Since then, however, the Interstate Commerce Commission, acting under the Railroad Revitalization and Regulatory Reform Act of 1976, has granted the railroad ten rate increases, bringing the price to over $18 per ton. With the support of the Department of Energy, San Antonio has taken the pricing issue to the federal courts. Other utilities have filed similar suits. To fight the increased price of coal, one Texas utility has tested South African coal and one in Florida is already buying 25 percent of its coal from Poland at a substantial saving. In the confusion, the United States, with the world's greatest coal reserves, is engaged in arranging to become dependent on foreign coal as a substitute for dependency on foreign oil.

We have had occasion to point out earlier that a straightforward, severe limitation on technological innovation in American industry now exists: lack of adequate cash flow. For a free enterprise approach to syn-fuel, would capital be available? Some argue that what hurts the return on investment in industry and makes for great reluctance to place capital at risk is the actuality or the fear of

*Congress has just created a "Synthetic Fuel Corporation" with initial funding of $20 billion.

government involvement. But would capital really be forthcoming from the private sector if the government could guarantee a hands-off approach to the problem?

Consider that a profit-seeking corporation choosing to make a major entry into synthetic fuel from coal would have to commit funds in the billions of dollars. A decade might pass before a return of any kind on that investment could be realized. Indeed, such a return might never eventuate, so great are the risks associated with unpredictable later government entry. The complete system needed for a syn-fuel industry based on coal involves a host of private and public organizations that are semiautonomous and not readily directed by any one body (not even the White House or Congress): landowners, mine operators, labor unions, railroads, power generating and water supply utilities, numerous engineering and manufacturing companies, county and state governments, and many, many agencies in the federal government that deal with prices, environment, safety, labor, antitrust, and transport, to name only a portion.

The problem of coal transportation by rail is a severe one that has to give pause to any free enterprise group contemplating investments in the billions of dollars. American railroads are already under a considerable amount of government control and the risk of inadequate capacity, in view of the railroads' financial situation, is thought by most observers to be a real one. Those speaking for the railroads usually are heard to say that they can improve their railbeds and provide enough freight cars to transport all the coal that can be mined. However, others have estimated it would take several tens of billions of dollars to expand capacity and upgrade aging roadbeds and equipment so that a major dependence on coal in the United States could be sound.

The environmental controls situation is equally dangerous for a strictly private investment in coal-based syn-fuel. One of the problems in trying to gauge what the government might do in the future is that the deleterious output of synthetic fuel plants has yet to be tested on an adequate scale. Syn-fuel from coal is surely uneconomic unless very high-capacity plants are assumed. A small experimental synthetic oil process has only to handle minute amounts of coal ash, but clogged filters could easily shut down a 100,000-barrel-a-day operation. Pollutants on a microscopic scale go unnoticed as they escape inadvertently. But in a full-size plant they can build up to be the equivalent of an ecological disaster. Because of the problem of measuring the effects of very tiny quantities, dangers to production workers are hardly noticeable until large-scale operations begin. Keeping in mind the eventual goal of high-volume output, it will not be easy for our imagined private entrepreneurs to be sanguine about pollution.

After 10 years and $10 billion put at risk, with a flow of synthetic fuel ready to market and the realizing of a return on the investment about to begin, the private entrepreneurial team has to reckon on the government's stepping in to clamp on price controls, or set more severe environmental standards, or file an antitrust case, or all three.

All these considerations suggest that the private sector is unlikely to develop a major synthetic fuel industry based on coal. In a realistically assessed social–political world, an attempt at a totally free enterprise approach to synthetic fuel offers a reward-to-risk ratio so low that embarking on a full-scale crash program would border on financial irresponsibility for even our largest corporations. It is probable the government will remain heavily involved in energy supply and will continue to have a $10 billion or more Energy Department. It will continue to be politically advantageous for those seeking elective office to win votes based upon the claims that the price of energy is too high because of excess profiteering by the private sector. Unless we invent something new, the most that can be expected (and which is already happening) is that exploratory private programs will be carried on in coal-based syn-fuels.

We had better look for some other approach to make realistic the option of coal syn-fuel in the United States on a substantial scale. In seeking that approach let us keep in mind the fundamental idea we have been emphasizing: the suitable arrangement for America is one that uses the government and the private sector in the right roles and missions. Government should take on those aspects it alone must accomplish. The private sector should do those tasks it is more capable of getting done than government. We shall now outline such an approach.

The DOE (Department of Energy), we propose, should announce that for the government's own use it wishes to purchase from the private sector synthetic liquid fuel made from coal. DOE will state the quantity and quality of the fuel it wishes to buy and the desired delivery schedule. It will offer a 10-year contract with a price adjustment factor for inflation during that period. The government, in choosing the competition winners, will give credit both for lowest price and earliest start on deliveries. The government will set standards as to safety and pollution that it will agree not to tighten without upward price adjustment. It will plan to buy the total — say, about three or four hundred thousand barrels of oil a day — from at least two sources. The government will provide immunity from antitrust if companies wish to create joint ventures to bid on the proposal. It will provide a proper cancellation fee if it wishes to cancel partway through the 10-year period. The government finally will announce a ceiling price above which no bid will be welcome.

It is submitted that if the government were to issue this request for proposals a number of firms of high competence and substantial financial backing would bid. The proposals submitted to meet the government's request would be sensible from the standpoint of the bidders because they otherwise would not submit them. Doubtless, the prices quoted in the proposals would be higher than existing petroleum prices but, from what we already know of the technology and economics, not so high as to rule out the program. Access to suitable land and water resources would be included in the proposals and the ultimate contracts. State or federal land would be made available at particular locations at stated price ranges as part of any deals closed.

Of course, approvals from various state and federal agencies would be involved. It has become apparent that environmental and safety regulations by the federal government and the states, counties, and cities have made the process of creating new facilities unacceptably long and complex, actually reaching the point where these regulatory procedures could prevent our ever constructing alternatives to imports of foreign oil. Some of the delays are merely the bureaucratic decision-making system at work. There are lengthy and repetitive judicial reviews as well as those of the regulatory agencies. Usually the standards put out by the government regulatory agencies change between the time a plant's construction has begun and the time it is finished. President Carter has proposed a so-called Energy Mobilization Board with unusual powers to accelerate this process, limit judicial review, and make overriding decisions on facilities it regards as critical. The legislation creating the syn-fuel program we have described could include provisions for the designated government agency (DOE) awarding the contracts to have certain of the powers envisaged for this (otherwise quite different) proposed Energy Mobilization Board.

If the Congress were to create the general investigatory agency on safety, health, and environmental protection (SHEP) as well as the decision boards we described earlier, these units would be active on the syn-fuel approach just outlined. SHEP would recommend standards to be met by the project. DOE's plans and SHEP's suggested conditions, if in conflict in any way, would be judged promptly by the decision board that would review the whole idea.

The government might also provide certain incentives by way of altering the allowed depreciation schedules for plants, making faster depreciation write-offs acceptable to the IRS. A principal reason why the price of synthetic fuels is high is the initial capital cost of creating the facilities. Over a substantial period of time this capital is returned through depreciation charges as they are absorbed in the sales price. As the years pass, $10-billion plant becomes a $5-billion plant and eventually a lower-costed one, on the accounting ledgers of the project. The annual return on shareholders' equity in the investment thus increases with time if one assumes prices and operating expenses stay constant. Obiously there is room here for some creative financing, accounting, and taxing patterns. For example, if the government were to allow accelerated depreciation for synthetic fuel plants (in the figuring of taxes for IRS purposes), the reported earnings would be low in the early years. The income taxes would also be low, improving the cash flow for the first part of the history of the facility. This is of unusual importance when we consider that private investment has to be especially fearful of changes in government policy that could impair financial performance in later years. It would be better from the standpoint of an investor for the facility to be written off rapidly, to cut cash payments of income tax to the government while the going is good, so to speak, in the early period when the capital invested at risk is highest and the income lowest.

If this program were created it would fully cover the requirement that America get started in a meaningful way on synthetic fuel from coal. It would set up the option to broaden the program later or to keep it as a lower-level program if other alternatives to the energy problem prove out better. The worst that could happen from the standpoint of the government is that if foreign oil did not rise enough in price during the contract period—and we would welcome that un-likely occurrence—the government would overpay somewhat for the synthetic oil it would have purchased for its own needs.

From the standpoint of appropriate roles for both the private sector and the government we notice that in this proposal the government is not involved in the details of the technology, where it has the least contribution to make. The government creates a minimum market for the output of the private sector. The government sets safety and environmental standards, which only the government can do, and which it must do in any case. The free enterprise industry makes the investment and takes a business risk, choosing the technology it favors. The technological company winning the competition will direct its program. The government will not.

9

salvation through conservation

One way to solve our energy problem, cutting American imports of foreign oil, ending fearsome shortage threats, and improving our trade balances, is to decrease greatly our waste of energy. Considering the rising price levels for oil and the certainty they will go even higher, we can realize an economic payoff if we lower needless energy dissipation—a benefit quite apart from the political advantages of lessened dependence on OPEC nations. Most who have studied the situation believe the political–economic environment for at least the next ten years will ensure that the petroleum supply and the energy flow from alternatives to petroleum will be too low to allow continuation of our present energy use habits. In contrast, they argue, conservation offers opportunities for a realistic lowering of requirements.

The conservation they speak of is not that attained by psychological pressure, by the government's urging users to wear woolen underwear in poorly heated buildings in the winter, or by hurting productivity when air-conditioning is outlawed in factories during the summer. Conservation means accelerating the inevitable mass national changeover to smaller and lighter cars of more modest but quite acceptable performance and comfort; speeding up the installation of improved urban transportation systems; adding heat insulation to homes and commercial buildings; applying new technology to increase efficiency of furnaces, boilers, and electrical apparatus; stopping construction of additional highways that have little to do with economic strength and growth, and using those funds instead to uplift the railroad system for greater efficiency; reshaping our cities to increase walking and bicycling; optimizing airline routes and schedules around fuel conservation; increasing R&D in agriculture to find ways to lower the need for petroleum-based fertilizers; installing computer scheduling of distribution to make the movement of goods more efficient; modifying industrial processes and practices to consume less energy.

Even information technology has a relationship to the saving of energy. A large fraction of petroleum consumption is for transporting people who merely want to exchange information with others. While the telephone helps do that, much more of all necessary communicating could be done through application of additional electronics. Sometimes the transfer needed could be accomplished merely by automatic delivery of computer data and at other times it could be done by face-to-face conferences through video telephony. Letters, memoranda, reports, and documents make up a large fraction of the physical transport of society. Almost all of such information could be put in the form of electronic signals and moved quickly and cheaply about the world. Electronic mail will save not only energy but time and reliability in delivery.

It has been estimated that over the next ten years a conscientious effort to incorporate all the changes just listed could save the nation hundreds of billions of dollars. This sum is figured by taking the price we would have had to pay for the oil otherwise used and subtracting the cost of incorporating the conservation steps (including R&D, design work, conversions, and installations). In 1978 a report issued by the Massachusetts Energy Office calculated that a $1.6-billion investment to insulate homes in New England would cut residential energy use 30 percent by 1985.* This investment, which would also create thousands of jobs, works out to over a billion dollars of annual saving at today's price for oil, an amount comparable with the one-time investment in insulation. Some analysts of the effect of conservation are even more optimistic than this one example suggests. They estimate that a thorough national conservation program, one still leaving our lifestyles essentially unchanged, could cut foreign oil imports to one-half by 1990. This would represent savings in the range of $300 billion to $400 billion for the decade.

This optimism is made to seem justified by the progress already achieved in energy conservation. In the past five years, while real industrial unit output overall has increased substantially, the use of

*Many figures on conservation can be found in Chapter 6 of the book *Energy Future*, edited by Robert Stabaugh and Daniel Yergin [New York: Random House, 1979].

energy has dropped: a 12 percent increase versus a 6 percent decrease. Half of our homeowners have added insulation to their homes and the annual growth rate in home electricity use has been halved, from 6 percent to 3 percent. The average new car purchased today will do 20 MPG, as against 14 MPG five years ago. The 1985 MPG performance should double the average MPG of cars of a decade earlier. None of this has been accidental; it has resulted from deliberate efforts to conserve energy.

Many corporations are now bragging in their annual reports about the success of their internal programs to conserve energy and the quantities being mentioned are of great significance. Typically the reports say that over the past five years, during which the quantity of output of product has gone up significantly, as has the square footage of the plants, the total energy used has decreased. The energy required to produce a finished ton of product has also dropped. The federal government established voluntary energy efficiency standards for the top fifty corporations in each of the ten most energy-intensive industries. Most of those companies are beating the suggested reduction of approximately 15 percent. J. C. Penney, which operates over 2,000 stores, has announced a decrease of 30 percent in energy use for heating and lighting through superior energy management.

It must be emphasized that such curtailments of energy use have involved the expenditure of substantial funds in detailed engineering effort and significant capital investment to alter processes. Without such technological developments as computerized systems for measurement and auditing of the operating parameters determining energy dissipation, the conservation results could not have been attained. (It is worth noting also that the funds expended could have been used to finance innovation for new products and increased productivity.)

Seattle has recently concluded that an electrical energy conservation program could enable it to hold off installing nuclear reactors until 1990. Nearby Portland has made it mandatory that a building cannot be sold unless it has had an energy audit and insulation installed as recommended. Portland also has banned automobiles from all but a few streets in its downtown area and has more than doubled the use of city buses. Counting everything, Portland expects to save between 10 and 15 million barrels of oil annually. A town in Ohio vented the heat from the computer rooms in its municipal building to heat the courtrooms above. A town in Colorado cycled its waste water onto nearby farms, cutting energy use for sewage treatment and fertilizer production. Community swimming pools are beginning to be covered at night. Hardware stores and energy-saver boutiques are beginning to offer a variety of gizmos to save energy: improved shower heads that use less hot water to yield the same pressure, coverage, and bathers' satisfaction; ceiling fans to push the hot air down toward the room occupants' feet; materials to plug heat leaks in baseboards and attics; transparent covers for windows.

By the government's recent deregulating of certain commercial airline activities, the airlines became privileged to operate more in a free market or competitive pattern. Under strict regulation, airlines competed with each other by adding service. This had the effect of driving down passenger load factors and the magnitude of passengers and freight carried per gallon of fuel. For 1979, the first full year of deregulation, this ratio grew by one-third over the earlier five-year average during regulation. As a direct result of better load factors, the largest energy usage improvement in the history of airline operations took place.

It would be strategically advantageous to the United States to be more independent as to energy sources and it would be very helpful in our relationship with other nations if we did not take so large a fraction of the world's oil as we now do. More than two-thirds of Europe's overall energy requirements have to be imported because those countries have no choice. The United States, with substantial domestic resources, consumes more than a third of the world's energy. Even disregarding economic advantage, therefore, conservation is important for its high political potential. Aside from this, it is foolishly wasteful if, having the option to reduce energy consumption and reap money-saving rewards by so doing, we fail to avail ourselves of such benefits because of ignorance or bad habits.

For example, the annual cost of improved insulation in many homes, when the total cost is spread over the life of the installation, is less than the annual expenses to generate the presently lost heat, although this may not have been true in years past when energy was much cheaper. In most commercial buildings, today's electronic instrumentation and computerized control make possible the same room-temperature comfort as earlier with the expenditure of less energy. The costs involved in the improved control are less than the costs of the energy saved. (A building can average 75 degrees by being held very close to that figure efficiently, or by oscillating continually between 65 and 85 degrees. The latter is both less efficient and less comfortable.) Similarly, electronic sensing devices with computerized control can direct shifts of air in buildings from the outside to the inside whenever it is advantageous, and the potential exists for cleaning the air so as to make the preconditioned air more useful. Finally, industrial processes mainly were designed in the past around optimizing costs when energy was cheap. If these processes are now redone, the costs to cover the redesign and modification are often less than the price of the incremental energy saved in one or two years.

In generating electricity by creating steam to drive steam turbines (which then turn electric alternators), the leftover or unused steam has usually been exhausted without recovering the heat energy residue it still contains. Many industrial processes require steam which could be used also to generate electric power. In general, industry needs both steam and electricity and they have been generated independently in the past. Cogeneration, the name for this combination of functions, offers a valuable source of energy conservation techniques that would be practical in a great number of situations. One study of cogeneration concluded that widespread use of the approach in the United States could yield more incremental energy than all the nation's existing nuclear reactors. The return on investment in equipment and installation to make the cogeneration possible, through realizing the value of recovered energy, is high wherever the technique fits the steam and electricity requirements of the particular facility.

In trying to conserve energy, it is interesting that sometimes we can go advantageously to an entirely different field of technology to accomplish a given task. For example, with electronic traffic control in our principal cities, it becomes possible to sense and anticipate the traffic flow and then time the traffic lights optimally. This reduces mandatory stops by vehicles at traffic signals, saves travel time, and lowers fuel consumption and air pollution. Quite apart from the matter of easing fuel supply problems, the value of the gasoline saved is greater than the cost of the electronic systems figured over the life span of the equipment.

To conserve energy when the issue is one of stopping waste and saving money at the same time, why does not the free market spark and fully implement the conservation potential? If there is genuine high return on investment to be realized by householders and the managers of commercial and industrial activities of the nation, why do they not act? Why assume, as many do, that to stop waste the government must step in to subsidize, pass new laws and police them, provide tax and other incentives, and in every other way force the citizenry to take steps that, if done without urging, would fatten their pocketbooks? If the government is going to be a financial backer, offering government money to help homeowners and private businesses decrease their costs of operation, the government will have to tax all of us to acquire the necessary funds. Then it will have to set up a bureaucracy to distribute the funds and direct the spending of them. All proposals for government aid will have to be reviewed by that agency and all implementations watched carefully to see that the aid is being properly employed. Government financing, whether in the form of low-interest loans, tax exemptions, or out-and-out cash grants for R&D or installations, has to come from the citizens. Why should the government collect money and then give it back (or what is left of it after the government's high administrative costs are met) in order to force savings sensible for the saver in the first place?

If the problem is only a matter of eventually stopping the self-controllable economic waste by individuals and industry, the government is not needed. It is actually wasteful to have the government involved. If insulating a home really will save the owner money, that fact soon will be made adequately believable through TV commercials. What if the homeowner lacks the necessary cash to invest in money-saving insulation? If the investment will come back with a good yield, a savings and loan association will gladly loan the money at commercial interest rates. They will hunt the homeowner down and say so, loudly and clearly. Should the government offer low-interest loans to cover insulation installations? Such loaning, below market rates, is a form of government interference that distorts the money market. This distortion denies to others legitimate borrowed funds and hurts many other aspects of the society. Some residences should be insulated fully; others should be left as is because of such details as the house's construction, age, location, heating system, the pattern of living in the house, the weather, and many others. There are too many factors for a government bureaucracy to work out general formulas and regulations to decide where government aid of what kind under what circumstances should be applied. The government's tampering with home insulation, to transfer funds from all taxpayers to a chosen few for government-directed purposes, is simply a way to use money less effectively and complicate practical choices by the end users.

Although this is true of home insulation, it is far more important to recognize the even more severe limits to the value of government entry into industrial processes and the heating and lighting of commercial buildings. If economic waste is the issue, we hardly want amateurs in government directing professionals whose mission in life is to obtain the best return on investment in their operations. Those who design and manage industrial processes are in the best position to decide how to alter the operations so that, in view of the increased cost of energy, the return on the investment will improve. Here, in particular, one sees readily the absurdity of creating a government bureaucracy, totally lacking the required expertise, to engage in negotiations, analysis, and approvals so that private industry can obtain government financial assistance to implement money-saving steps that should be taken anyway.

Consider some tradeoffs the free market should readily handle in pitting conservation against energy costs. A good example is the private enterprise investment going on now by several electrical equipment manufacturers to develop energy-saving technology in the operation of electric motors. Engineers generally prefer to specify AC motors because the alternating current employed is the form in which electricity is directly generated and furnished by the power utilities. Speed control has in the past been a problem with AC motors but, with energy very cheap, the practical answer has been to use mechanical methods to draw off (and throw away) power and thus slow down the motor to the desired speed. New semiconductor technology now can be applied to control motor RPM without this waste, while maintaining an economical price for the equipment. The energy decrease ranges from 15 to 30 percent so the annual dollar saving is usually more than enough to justify investment by the user to purchase the new product and by its producer to create it. No government participation is needed or sensible.

Another example illustrates how availability of savings to the consumer and the free market should cooperate well, as the public becomes more aware of the high cost of energy. In the past, light bulbs have been regarded by the purchaser as relatively cheap and the price of electricity to power them very modest. The consumer has thus favored low-priced bulbs at the store. Now, new technology makes possible light sources that last longer and deliver five or six times as much light for the same wattage dissipation. These bulbs are more expensive in initial purchase cost. However, over the life cycle of the bulb the user will save more in energy costs than the initial purchase increment (savings are estimated between $20 and $40 per bulb). As the price of electricity rises, the bulbs will be put on the market with no government subsidies for light bulb purchases or R&D funds needed to develop them.

Energy conservation to eliminate economic waste is a natural for the free market. The only contribution the government has to make here, and it is a vital one, is to allow the free market to operate. The government must only set the rules as to safety, health, and environmental pollution. Whether a private operation is left unchanged or modified to seek the money-saving potential of energy conservation, the activity will still have to meet standards set and policed by the government. The free market is a route to conservation when economic factors control, provided the government stops price subsidization and allocation controls. A near consensus exists among energy experts and economists, for example, that the abandonment of all price controls on petroleum is an essential ingredient of sound conservation policy. These people believe that only exposure to realistic prices can convince American consumers that the era of cheap, abundantly available fuel is over. The marketplace is far superior to any amount of presidential rhetoric to encourage conservation and inhibit wasteful consumption. When it comes to economic tradeoffs alone, ignoring for the moment the political dimensions, the market can match up supply and demand while encouraging the highest possible supply because private enterprise will have the incentives both to find more petroleum and to develop substitutes.

Our present national MPG average for cars is less than 14, while in Europe this performance figure is over 20. Since the Europeans pay more than twice what we do for a gallon of gasoline, it is hard for them to believe we are realistic about stopping our squandering of energy resources. We take more than our share in competition with them, driving up the world price. We talk conservation but meanwhile subsidize a price below the true market value of the gasoline. The price rise in gasoline during calendar year 1979 was accompanied by a drop in consumption of 8 percent compared with the preceding year's figure. Recent analyses have suggested that a doubling of gasoline prices would mean a decline in consumption of another 20 percent.

Having mandated prices on gasoline below the rest of the world's retail market price, the government has had to accompany this approach with some sort of allocation scheme to determine who is to end up short. A White House economist has been quoted as saying, "Running any allocation system is like driving while looking in the rear view mirror. It works fine while you are on a straight road." And, we might add, with no obstacles ahead. A bureaucracy running an allocation scheme is very likely to have the gasoline where the cars are not and the heating oil far from the location of the cold spell, and to be busy responding to the latest political screams. When farmers get vocal about not having sufficient diesel fuel, their representatives get a sympathetic hearing and more diesel fuel is then set aside for tractors. The truckers next become angry and exert their political pressures. As long as there is political worry about the effect of higher-priced oil and gasoline on voter support, dropping controls is not tempting to a politician.

Holding down the price of gasoline is not even a good way of assisting the poor. They have problems with buying everything — food, housing, clothing, and medical care, as well as automobile transportation. Why especially encourage them to buy gasoline with their limited funds? If we want to help the poor there are direct ways to give them more to spend on what they themselves will deem to be their most urgent need, which, in individual cases, may not be gasoline.

For political reasons, the government is unlikely to stay out of the energy pricing picture. The potential of conservation goes far beyond matters that can be classed as related to waste or economic gain. Reducing American dependence on foreign oil has political and social advantages in addition to economic ones. If we use less of the available world oil, we will cut down world tensions that are bound to grow as the energy supply problem increasingly affects the economies of all nations. If we reduce our demand for oil through conservation, we shall have less need for accelerated development of nuclear power or synthetic fuels from coal or shale. If we can be less nervous about these options, with their safety and environmental pollution fears, life will be easier in America. Because of the importance of these political–social issues, we must strive to conserve energy beyond the economic gain maximization level which a free market, if it existed, would achieve.

At the moment when departure from an economic optimum becomes a significant factor in national objectives, the free market alone will not provide the broader optimum we seek as an overall economic–political–social balance. To devise a program to cut energy consumption beyond the economic optimum point, what should we do? Again, we should ask about the best roles for government and the private sector. We should avoid assigning to the government any tasks it cannot do well. This includes decision making on that class of details where the government cannot be expected to have the necessary information and competence. We should allow maximum private discretion about conservation by individual citizens and corporations, and avoid the enormous penalties of highly bureaucratic rules, regulations, and policing. Where a myriad of local, individually tailored actions are to be dealt with, the federal government has an excellent chance of doing the wrong things at least as often as the right ones. For those situations, a free market pricing system will do a better job of allocations than government involvement. Let us take specific examples.

Some like it hot and some like it cold. Some like to travel and others care little about it. Mrs. Jones may choose to have her home at a low temperature and wear a sweater to save energy, but may find it very important to take a regular airplane trip, which uses energy, to visit her children. Mr. Johnson can use a carpool to save gas while his neighbor who works irregular hours cannot, though he rarely travels out of town. The retired Mr. Smith has arthritis and does not move around very much. He wants it warm in his house and is willing to use his car less. The price of energy is a factor for these individuals but clearly there are other considerations as well. If the government sets out to create rigid conditions that all citizens must meet, denying them the privilege of making personal value judgments, government actions will be arbitrary and unfair, or will be seen as such, and will cause vast non-compliance even if the government polices heavily (and expensively) trying to ensure adherence to the rules.

Granted, if it is considered in the interest of the nation that we cut down foreign oil imports beyond the economic balance level set by a free market, then government influence toward this end is essential. Still, the government entry should be reserved for those areas of the nation's endeavors that are large-scale in impact and where private investments and consumer actions in the free market are not likely to come close enough to achieving the desired conservation ends.

As an example, Dow Corning Corporation, without government involvement, is building a $30-million plant to provide steam and electricity for its Midland, Michigan, operations. The plant will burn wood. Dow Corning is switching from natural gas to wood with an estimated saving of 30 to 40 percent on fuel and $5 million in capital costs. The U.S. Forest Service estimates* that thousands of plants the size of Dow Corning's could be fueled by current U.S. forest yields without cutting into supplies for building materials and paper products. However, the savings to be realized are a function of details. (For instance, Midland is surrounded by millions of acres of forest.) Dow Corning is in a much better position and more competent than a government agency to figure out whether the decision to burn wood is of economic advantage.

As another example, substitution of coal for oil to fuel the large boilers for electric power generation requires large capital investments by the utilities to effect the changeover. How far the nation can go in conserving petroleum by substituting coal in electric utilities' boilers depends partially on technological advances in every aspect of the mining, transporting, and utilizing of coal. If we could make

*Business Week, February 18, 1980, page 110L.

rapid progress in learning how to burn coal more cleanly and to remove the remainder of the pollutants more economically, coal would be increased in its attractiveness. A better chance of good financial return on greater use of coal would exist, and more private investment in this option would be made. However, the nation may need a magnitude of petroleum conservation—and substitution of coal for petroleum—going beyond the point of optimum economic tradeoff. The free market then will not carry the nation as far as is desirable in developing the technology for further use of coal. Government intervention to some extent is indicated.

The utilities hardly operate in what may be called a totally free market situation. The electric power utilities' rates and their depreciation schedules are government controlled. The policy of keeping allowed rates low, together with out-of-date IRS depreciation rules, have caused American utilities to be strapped for capital for expansion and maintenance. If we add to this a requirement for a changeover to coal, with very stringent pollution control standards in the burning of it, the utilities' managements and shareholders are not in a position to interpret what they see ahead as representing good returns on invested capital. They will have great difficulty raising needed funds. Modified government-allowed depreciation policies, tax incentives, and government guarantees of loans are the kinds of steps the utilities might need, and these measures only the government can arrange.

The government might provide a subsidy to a utility to substitute coal for petroleum, increasing the subsidy as the stack is made cleaner. This would impel the utility to spend some of the initial subsidy funds in an effort to improve coal usage from the standpoint of pollution. The motivation of the private sector would cause the investment of the subsidy funds granted by the government to be spent in technological advance efforts yielding the greatest economic return. In this instance, the government investment would move the utility above and beyond the economic tradeoff level. It would be a sensible act by government because it would create economic incentives for the utility to accelerate the changeover to coal. This is admittedly interference with the free market. However, because of the (permanent) interference that is already taking place, and because of the importance to the nation of the political–social aspects in addition to the economic ones, it is sound for the government to participate in some such way.

Still another example brings out the need for selecting the right respective missions for the private sector and the government so as to achieve the desired level of conservation. We could conserve more petroleum if we could move the nation more rapidly toward higher-MPG automobiles. The free market has not been fully employed so far to achieve this because the government has not allowed free market pricing of fuel. In the past few years the government has been regulating the MPG of new cars and propagandizing to cut down gasoline consumption by American automobile drivers — while at the same time trying to keep the gasoline prices low through price controls, thus encouraging gasoline usage. Can we do better than engage in this set of government-edicted inconsistencies to convert America faster to lower gas consumption autos? If, by the rubbing of an Aladdin's lamp, 100 million American cars could be changed immediately to smaller, lightweight ones with twice the MPG figure, we would realize a corresponding huge reduction in overall petroleum consumption and in the size of oil imports. This full changeover presumably will take place eventually. What are the best roles for the government and the free market, to effect this changeover as soon as possible?

If free market pricing of gasoline were adequate we would have no problem. If the price of gasoline rose enough, the demand for low-consumption cars would increase and all foreign or domestic automobile manufacturers selling in the United States would work as hard as they could to capture the exploding market for smaller cars. Existing high gas consumption vehicles would drop in price and the older ones would be junked and converted to raw materials. Unfortunately, the increasing of gasoline prices in the free market will not by itself move the nation to conserve as rapidly as desired. People hesitate to give up existing cars before they have depreciated adequately from use, because of the large trade-in loss. That loss has to be pitted by owners against the saving in gasoline when they buy a new small car as a substitute. An average person driving 15,000 miles per year in a car that does 15 MPG at $1 a gallon uses 1,000 gallons of gasoline and spends $1,000 annually. Add a big $1 to the price of gasoline and this adds $1,000 to the expenses of operating that car for a year. The owner could save $1,000 on fuel by buying a car that gets 30 MPG. But what of the trade-in value of the present, low-MPG car? As far as the pocketbook is concerned, the owner should keep the larger car if a loss of more than $1,000 per year would result from an early trade-in.

If for purposes other than the economic tradeoff a more rapid changeover of the nation's cars is desired, some government action is indicated. But a wide variety of possibilities exist for that action. One that we should rule out right away nevertheless has received a certain amount of support. It is that the government should spend the taxpayers' dollars to fund the development of a high-MPG car. Surely the automotive industry, both domestic and foreign based, has adequate incentive to develop such a car already. Investing private funds innovatively, the industry stands to reap the reward of a favored position in the marketplace. If the existing, highly competitive automotive industry does not know or cannot learn how to develop the most suitable car, it is not clear how a government bureaucracy that lacks the detailed expertise can do better. We can save ourselves tax money by not sponsoring government efforts along such lines.

Another approach is to place a high sales tax on gasoline purchases at the pump, say, 50 cents or one dollar per gallon. The government could use the revenues collected to expand and improve public urban transportation or lower income taxes by the same amount. The income tax relief might be biased to cut taxes at the lowest levels of income where the penalties of higher gasoline prices will be most felt. Obviously, no end of proposals can be made of ways for the government to impact the free market so as to create a lowered demand for gasoline than the free market alone would set. One approach the government has already employed is to lay down strict rules for the MPG performance of cars admitted into the American market. With these laws in effect, car manufacturers have been moving in the direction of higher-MPG cars. This promotes conservation of fuel beyond market influences based on cost of fuel.

However, there is much more to it. If the government is too demanding in MPG improvements, it can make it impossible for some auto manufacturers to stay in business. Full-scale redesigns of cars and retooling of plants, paralleled by severe standards on pollution and safety, require enormous investments by the automobile manufacturers, while they meanwhile are realizing less positive cash flow from present operations. Chrysler has already found it cannot keep up on its own. Ford is having a difficult time. In the United States, only General Motors may be financially strong enough to handle the government requirements to move rapidly toward more gas-stingy automobiles, if the demands escalate further. Since the foreign producers have been making smaller cars right along, the American changeover has been a bonanza to them. That may be temporary but it should be a factor in government policy formation and may not have been.

The automobile represents so large a chunk of the American economy that the changeover to a new-car inventory of 100 million smaller cars, which the nation should ideally have for the period ahead, must be handled wisely and carefully. Because of the energy problem, we want this massive inventory turnover to happen quickly. However, if we are not careful, we can create a combination of mass unemployment in the American automotive industry, negative balance-of-payments effects as we import more cars, and economic chaos, even as we succeed in cutting the amount of imported oil. The government must manage most of the complex system of tradeoffs involved here because the balance desired cannot all come out of the free market. However, the government should not use simplistic approaches. Instead, it should subject the critical automobile situation to detailed econometric study, looking at all of the interfaces of oil consumption, safety, air pollution, unemployment, and impact on international trade balances. It should seek the most sensible middle course in its demands for redesign of cars. Rather than fussing with gasoline price controls, oil companies' profits, and allocations of gasoline, when the free market could do a better job, the government's role should be to seek to anticipate the interfaces of the automobile issue with national goals. For instance, it may be that the policy of steadily making pollution minima more severe should be slowed somewhat in the interests of greater MPG performance.

There is an enormous market in the United States in the next decade for automobiles, because, in addition to the normal annual requirements for cars, the superimposed, large-scale changeover just described is certainly going to occur. The market potential of this changeover means private capital will be attracted to the automotive field — if the government strives for the greatest of clarity and stability and the best balance in the requirements it sets down. To do otherwise can turn away investment capital needed to complete the changeover. Because of the importance of the automobile to the general economy and the energy situation, the danger is that the government will get into the act in an overly political, hit-or-miss way and make the situation worse from all standpoints. As the automobile industry suffers, and with it the entire nation's economic stability, and as energy problems worsen, the tendency will be for the government to enter into the automobile situation ever more deeply. Automobile production, like energy, could become then a multidimensional political mess with severe economic and social consequences.

The automobile has become for us an outstanding example of the limitations of both government and free enterprise. Those who think the free market alone should totally govern what happens to the automobile in America are neglecting the political and social factors that go beyond the interests of the automobile industry and the individual drivers of cars. Likewise, those who think the government should handle everything will lead us astray if they fail to recognize the continuing importance of the free market aspects of the automotive field. The automobile illustrates well the problem of technology–society interactions, because it shows that a hybrid of free enterprise and government control is needed. It shows that the problem is to find the right pattern of cooperation of the private sector with the government. The automobile situation further shows that we have not yet learned how to do this organizing very well.

10

urban transportation in slow motion

The best use of science and technology on behalf of American society will result from a rational combination of free enterprise and government control. Prudent choice of the role and mission for each sector will depend upon the specific area of endeavor, the problem to be solved, the activity to be implemented. We have discussed this hybrid pattern for the organization of the nation's way of managing some key aspects of energy. Let us now consider other important categories of the nation's science and technology effort.

We start with a look at urban mass transport. Almost every American city of reasonable size could gain by installing first-class public transportation. But it has to be the right system, with the technological approach carefully selected and matched to each area's pattern, industry spread, population and employment distribution, health care and educational facilities layout, and relation to suburban regions. The mass transport must connect solidly into the specific city's economy, social conditions, and lifestyle. Assuming sound application of the correct technology to the city's needs, the economic gains could be prodigious. Perhaps half our population would save hours per person each week getting to and from work. They would travel at lower expense, consume less energy, diminish air pollution, and experience fewer accidents. If America is moving toward the realization of such improved urban mass transit it is by very slow motion.

Is it intelligent for a private American technological corporation to invest its resources with the objective of developing, then selling, and finally earning a return from modern mass transit systems for our cities? Consider Los Angeles as an example. Imagine that some adventuresome company had devoted several years of effort to fitting technology to L.A.'s social and economic needs and had designed an appropriate multibillion-dollar transit system. Then further assume, as we must to envision a realistic system, that all the R&D, design work and approvals (vehicles, controls, communications, energy sourcing and distribution, safety system, passenger stations, repair and maintenance facilities, rights-of-way legalities, environmental reports, etc.) had been completed. All this would have involved an expenditure of many millions of dollars and would have been only a beginning.

Much of this effort would have had to be repeated for other cities because of the importance of bringing prices down by realizing as much commonality as possible in equipment and basic system architecture among different applications, and by selling multiple installations. Next, the private concern or its subcontractors would have had to design, build, and tool up several manufacturing plants to start producing the required hardware and refining it by full-scale testing. Parts would have to be manufactured in some quantity to be sure of their costs and reliability before committing them to contracts for shipment and installation.

If the company had gone this far—it would by now have invested several hundred million dollars—and had done everything perfectly, would it then have succeeded in selling a system to Los Angeles? Probably not. Who exactly is this customer we have labeled Los Angeles? What does Los Angeles truly want? How are its requirements being decided? Can Los Angeles raise the procurement funds? Hundreds of separate (and often quarreling, apathetic, or purely self-seeking) groups in L.A., both private and governmental, are involved in the answers to these questions. Potentially beneficial as the creative application of effective transportation technology might be to the city, the market in Los Angeles is not formed. The reward-to-risk ratio would be unacceptably low for any private corporation electing to develop this field. The start-up cost would be huge. Even assuming technical success eventually, the time to payoff would be too long and the city or county government's eventual regulating of the pricing of the service, fundamental to the promise of return on the private investment, would be too unpredictable.

Clearly no company is going to do what we just described: invest huge funds with the hope of a favorable economic return in the design and building of entire systems for urban transportation. Of course, a city could decide to go ahead on its own, instead of waiting for free enterprise to anticipate potential profits and act on that promise. A municipality or county could contract with a competent private firm to perform systems engineering and architecture on the transportation network so that subsystem hardware (vehicles, computer controls, passenger stations, etc.) could be specified and put out for bids to industry. The system then would get built through a hybrid of government and private involvement. This is the way the Washington (D.C.) subway was constructed.

The Washington subway is often pointed to as an example of how not to do things in urban mass transportation. In the first place, the area has a substantially higher per-capita income than the nation's average, yet billions of dollars of federal funds, money drawn in taxes from the citizens of the nation at large, were spent to benefit the residents of that one region. There were numerous delays in construction, overruns on original estimates, doubtfully justified add-ons, and many design changes resulting from newly mandated governmental regulations. All this caused the cost to more than double during construction. The choices of routes and locations of stations (and even the selection of technology and overall systems approaches to match up engineering with requirements) often were more political than logical or objective.

Such problems can never be totally eliminated, but good organization could have minimized them. At any rate, the presence of difficulties should not blind us to the value of good urban mass transit. Our situation in America is not greatly different from other parts of the world, such as Europe, where it is recognized that urban living is a permanent phenomenon and that effective public transportation makes for a higher quality of life as well as providing for economies and easing of energy, safety, health, and environmental problems.

Another presently practiced arrangement for urban transport is worth mentioning. Some urban transportation requirements are being handled today by the cities' merely turning the problem over to a local bus utility, which buys buses from a private corporation such as General Motors. Presumably, as with its cars and trucks, GM tries to anticipate the coming need for buses and designs the kinds of buses that will meet the need. It spends some of its available risk capital to place itself in a position to win competitive bids. To help its sales along, it tries to persuade potential customers (the cities' transportation units) that the buses it offers for sale will create benefits greater than the costs of adding them to the transportation system.

But is this enough? Are these kinds of relationships between the cities and the technological industry the best we can arrange so as to put technology to work? Is the present situation satisfactory? Obviously not, or most cities would now have excellent transportation systems. In fact, essentially none do. Something is missing. We should be able to do better.

The potential market is certainly big enough to justify attention. About half the people of the United States live in cities with poor public transportation systems. The value of the time and energy waste which this inadequacy of transportation represents is enormous. A typical resident of Los Angeles, because of the layout of the city, drives 10 miles to work. With the existing meager public transportation, there is practically no choice but to use a private car. The 10 miles of displacement of car and body are accomplished at an average speed closer to 10 miles per hour than to the 100 miles per hour of which the automobile is capable. The worker drives alone and leaves the car standing in a parking lot all day, meanwhile getting no return for this investment and giving rise to an investment (that in the end must be paid for) in the parking lot. The auto pollutes the air as it is driven both ways in heavy traffic; it uses much gasoline when supply is an increasing problem, and it lowers the driver's life expectancy. In view of the miserable traffic strains, time is lost in the required calming down after arrival at work and there may be nastiness toward the family after arrival at home. Aside from the financial costs, the air pollution, the wear and tear on the nervous system, and the waste of energy, the driver accomplishes less for a given time, portal to portal, than if a good transportation system were available. Productivity is a major issue for the United States in economic competitiveness and standard of living. Surely, urban transportation improvement is the equivalent of productivity gains.

The picture we have just described can be duplicated for the missions of various family members as they go shopping, visit the physician, or commute to school or the sports arena. We should not ask for perfection in urban transportation, but modern technology should make it possible to do a better job of transporting people and things around a city. Los Angeles is among the worst cities as to public transportation, but anyone who travels in the United States knows it is not unique.

Cannot the private sector take a bigger step in developing and meeting this market potential? True, we described how unlikely it would be for a company or a group of investors to research the market and act on it to the extent of developing entire systems for sale. That is only one, extreme approach. Private industry could justify taking a bigger risk than just selling buses, if superior, practical means were apparent. The government could help create this better approach by accomplishing those tasks and making those decisions that only government — city, state, or national — can make.

To some there is an easy answer to the challenge: simply set up a huge federal government budget for public transport and dole money out to the cities. Equipped with these funds, the cities would then arrange to have modern urban transportation systems designed, equipped, and installed. We can judge the price it would be worth paying by returning to our postulate that about 100 million people in the United States live in cities where improved public transportation would pay off. Let us say that half of these, about 50 million, require daily transportation. If a city dweller gets around in a private car, the individual investment required to make this possible is from $5000 to $10,000 and the operating cost of the auto is over $2000 per year. Assume next that urban transportation might provide the equivalent service when shared with others at substantially less cost. Use, say, $2000 per person for this lower cost of the installation and equipment and $500 per year for the costs to the rider. Multiplying the first figure by 50 million users leads us to a national, first-time installation cost of $100 billion. If we further assume it will take 10 years to get all of the construction accomplished and operations going, we are talking about an investment of around $10 billion per year to complete the $100 billion total. If each rider spends on the average of $500 per year for the service, that yields a figure of $25 billion for annual revenue, some of which could be allocated to amortize and pay interest charges on the invested capital. As in most calculations of this kind, we are probably optimistic in money and time. However, these figures suggest a feasible combination of required investment capital and savings to the user. The total funds required do not appear out of line to achieve the anticipated energy savings, reduced pollution, time savings for the individual, and substantial contribution toward freedom from imported oil. The system could be financed with bonds bought by the citizens whose transportation savings would yield them funds for investment.

All these benefits and others are contingent upon accomplishment of a top-grade job of systems design so that the technology represented by the transportation system fits well into the cities' patterns. As previously discussed, the citizens must want to use the system once it is built. But how is this quality of effort to be reached? Let us describe the skeleton of a scheme.

In this proposal the federal government would furnish some of the funding, the community to be serviced by each installation would supply the rest. An urban transportation project office would be formed within the federal government — the Department of Transportation would appear to be the logical place — to exert executive control over the funds and the contracting to expend them. It would set program objectives as a foundation for the individual contract efforts to follow. Through a competition, the project office would select a team from private industry to provide systems engineering, overall architecture, and technical direction. This team, enhanced through appropriate subcontracts, would contain the requisite engineering and economics expertise in all the disciplines pertinent to the task. The government project office also would employ a small, independent overseers' board of outstanding citizens to help it watch the progress and confirm that an exceptionally competent and thorough job is continually being done by the selected systems engineering contractor. That contractor would set up field teams in each of an initial group of, say, 10 cities invited to participate, choosing them to provide variety in size, weather, nature of terrain, population spread, industry concentration, and specific needs.

Each city chosen as a candidate for a new transportation system would be allowed some privileges, but it would be obliged to take on certain responsibilities. If it is accepted for participation, a city would receive funds to enable it to put together a local team of its own to work out criteria so that the transportation system will meet the city's needs as the city sees them. In addition, each city would be expected to provide its own matching funds for perhaps half of the total required for the eventual system installation. The remaining funds would come from the federal government's project office.

The efforts of the project office's systems engineering contractor would include initial research and development on alternatives, the obtaining of experimental data as may be required to lay a basis for design, the making of economic tradeoff studies on optional approaches to the many detailed problems, and finally the supervision of contracts for development of equipment for installation. The equipment contractors would receive their contracts and funds directly from the project office through competitions. The rules used for contract administration would be similar to DOD's ICBM program or NASA's Apollo project.

Before ground would be broken in their towns, the communities would have a veto power. They could ask for changes in certain aspects of the design as it gets developed, of course. If they wished to stop the program in their area and bow out, they would have the privilege of doing so. In that instance, they would not have to invest their portion of the funds. On the other hand, neither would they get the transportation system they otherwise might have had. It is presumed the federal government's project office would want to go ahead with plans once they are worked out for the various cities.

There would be adequate motivation to try to arrive at a system that pleases the city. But if the project could not be finalized by negotiating and cooperating with the city, the federal government's project office simply would put its funds elsewhere.

Of course, all installations would have to meet sound standards on safety, health, and environmental protection. If the organizational entity described earlier (SHEP) were already brought into existence it would include public transportation in its investigations. SHEP would observe what the federal government's urban transportation project office and the cities were considering and planning to implement and would make its independent studies and recommendations as to standards and safeguards. If there were agreement, all would be well and good. If not, the matter of difference would go to a decision board as previously described in the chapters on energy. Here the two opposing parties before the board would be government units, one (the project office) striving to equip the nation with effective, economical public transportation in our cities, the other (SHEP) trying to protect the public against deleterious effects. The decision board would weigh the tradeoffs and have the authority to set the balance.

The systems engineering contractor would be expected to make a great effort to achieve equipment commonality among cities, where sameness would make sense. Only by such an effort, accomplished with great competence, can the national costs be kept down. The approach need not be inflexible or standardized, for example, monorails to be used everywhere or nowhere, or subways ruled out for each city if not believed to be the right pattern for all. Adaptation to the variations in the nature of the cities could coexist with considerable commonality, and significant economic advantage could result from harmonious, balanced optimization.

The job of construction and installation would be given to winning contractors, each chosen in a separate competition for the individual municipality. The installation contractor would operate under the supervision of the systems engineering contractor but would be funded by the federal government's project office.

As rapidly as possible, the systems would be expanded to more and more areas, and improvements would be introduced into each installation as time goes on. The policy would be to have at least two sources for every key component (vehicles, controls, computers) required in the system, so that the national urban transportation system would not depend on the performance of a single contractor. In general, it would be expected that teams of contractors would be formed to take on various jobs, with a lead contractor and a group of associate contractors. This practice is common in major construction engineering projects, both civilian and military.

The program described is quite different from past practice in which the federal government simply doled funds out to the cities. In this proposal, the federal government, through the project office just sketched, would be in the management business. It would have overall executive control and systems responsibility even though it would exercise these functions largely through hiring private expertise. This plan would require a long-term commitment by the Congress. A congressional committee would oversee the effort, receiving periodic reports from the project office as the program progresses.

We would expect a generous amount of political maneuvering by cities and states, pressing the federal government's project office for priorities and attention and for modifications of designs to go with local desires. We would look for no miracles. No organization inventable could eliminate the selfish interest pressures and noise making which are part of America. However, the organization of the effort we have pictured would provide the right roles and responsibilities for the government and the private sector. The relationship between them would assign to each that which it can do best and must do well for effective results. For just this reason, the proposed organization would possess a good chance of overall success. The installed transportation systems should represent a harmonious combination of usability, reliability, and safe performance. They should conserve energy and exhibit economical installation and operating costs.

11

food leadership by America

Let us shift now from the city to the country. In an earlier chapter, we cited food and nutrition as associated areas of endeavor where scientific research and technological development offer potential benefits to the society so significant we would not want to be denied them through neglect or misunderstanding. What should a sound program for the nation encompass in these fields? What are the appropriate roles for free enterprise and government?

The success of the United States system of food production hardly needs to be proved. Americans on the whole are being fed relatively cheaply and well and we produce large quantities of crucially needed foodstuffs for export to the world. This is accomplished using only a small percentage of our total labor force. We possess $100,000 farm machines that can harvest $100,000 worth of soybeans in a day. The machines we now employ on our farms number in the hundreds of thousands. If they were to join up for the task, they could harvest all of Iowa in 24 hours. Without the machines, bringing in the United States crop, which was worth $60 billion in 1979, would have required 30 million farmhands and 60 million mules. However, the unrealized possibilities, looked at as meeting either a world need or an American opportunity, go well beyond anything we have accomplished to date.

The present high stature of America in food production suggests a reasonably well formed pattern of activity, with both government and free enterprise making contributions. The food in America is grown, processed, packaged, delivered, researched, and regulated through an extensive participative network of private industry, gov-

185

ernment agencies, and educational institutions. What needs changing in mission responsibilities lies almost entirely in the government category. The private sector already plays the right role, but it could accomplish much more in that role. If the government's assignment were bettered, the climate would then exist for a greater contribution from the free enterprise sector. New opportunities would become evident and human and capital resources would move in to develop the free enterprise potential.

Before listing what the government should do in food and nutrition, we would do well to say what the government should stop doing. In the future, gradually but determinedly, it should lower its interference with the free market for food by diminishing its price management efforts and curbing subsidies. Let this recommendation not be equated with a lack of appreciation for the political and social sides of the food production industry. Nor is it suggested that the American government allow the nation a stronger dependence on the marketplace in food because free enterprise is believed to be the total answer to all aspects of food growing, processing, and delivery. Very important, highly critical roles for government will remain, including some the government is not now filling. However, in the decades ahead, the expanding market for American agricultural products, considering the entire world's needs, should improve returns on investment. This will make the need for politically motivated, short-term government entry—to come to the rescue of beleaguered private farms—far less necessary. Especially will this be true if the government provides the proper leadership in a national policy and strategy in food and nutrition, a most essential task and one which only the government can perform.

The government needs to be the agent for action in food and nutrition in four categories: (1) national policy and strategy, (2) basic research, (3) regulation for safety, health, and environmental protection, and (4) arranging for the availability of water, land, and energy.

By policy and strategy we mean the handling of the overall systems problem, already described in an earlier chapter. Food and nutrition for America, and America's role as a necessary supplier to the rest of the world, are a complex interactive array containing probably more pertinent and more influential parameters than any other area of human science and technology. For most nations, food supply is already a serious matter; in less than a decade it will probably be recognized as the No. 1 world problem. The prediction is that malnutrition in the world will worsen. The per-capita world production of food appears to be on a decreasing curve (temporary ups and downs not counted). Throughout the world, agricultural output is generally endangered by improper irrigation, by the impoverishment and erosion of soil and by overgrazing and overuse of marginal land. Hence, food will be a dominant international political issue. The United States government, not the free enterprise sector, must lead in relating food supply to political policy and must direct our tactical reactions to international conditions. As both energy and food shortages worsen, more and more nations will find they have to pit energy needs against food needs. In the United States, competition for land and water allocations will develop between energy producers and agriculturalists. Environmental interactions will proliferate also.

Among the examples: it is now evident that standards for allowable air pollution from synthetic fuels based on coal or from the burning of any fossil fuels should be set only after considering their impact on agriculture. Oil from shale requires enormous amounts of water; so does agriculture. A direct energy–agriculture interface occurs if biomass looms as a serious source of energy. Yet another example: the overall federal budget for R&D is bound to be limited, and competition will arise between spending the available funds for food and nutrition research or, alternatively, for energy, oceans, military, medical, and other research areas. Finally, however wise it may be to foster a free market in food products in the United States, the moment we add the exporting factor we are forced to depart from total reliance on the free market and the government must make export policies its business.

No matter how large our surplus in food, it alone will not satisfy the world's needs, whether of developed or underdeveloped nations. Thus the government will be forced to do some export allocating, playing favorites here and there, seeking to obtain the maximum contribution to our balance of payments in some areas, and generally using access to our food supply as a factor in diplomacy, security, and economic competitiveness. An example is the exceptionally large, but not overly stable, Soviet Union import requirements for grain. Soviet purchases can greatly affect prices in the world market and many American citizens and high government leaders will feel our exports to the USSR must be related to U.S.–Soviet cooperation or differences on other fronts. As another example, our export strategy has to include concern and special arrangements for the highly populated but impoverished nations. It is almost certain we shall always be providing some form of aid to them and that will affect the domestic and worldwide price structure for food.

To maintain and even enhance our leadership in food production is of prime importance to the United States. The route to this is to arrange not only for ample supply in the short term but also for a parallel superiority in the science and technology that provide the foundation for food production. Valuable know-how is, of itself, an item of advantageous export potential. Our technology will often be as desired by the rest of the world as the product flow that results from it. Our government will have to include technology transfer and research and development aid to other nations as part of our strategic and tactical policies in food and nutrition.

The second essential function we listed earlier for government is the sponsoring of research. Large private R&D activities are carried on in some parts of the food industry. However, farmers and food processors are engaged in very little basic research for reasons similar to those that cause the manufacturing industry of the nation to perform only modest pure science investigations. Research efforts that go to the very fundamentals of the biological and chemical characteristics of plants and animals do not easily lead to proprietary food product opportunities. Such research provides neither immediate edges over competitors nor short-range high return on the investment.

Long-range agricultural and nutritional research has been accomplished traditionally through a unique federal–state partnership, the Department of Agriculture's sponsorship of land grant universities, and some pass-throughs of pertinent scientific principles from the large government-sponsored research programs in biomedicine. These programs have two important shortcomings. First, the scale and breadth of research do not match up to the opportunity. Second, with the American agricultural system having by now reduced the lag between technological advances and their full application on the farm, our agricultural approaches now are seen by the experts as having almost used up the yield-increasing potential of past R&D. This nearly absorbed backlog of advantageous ideas for action stemmed from basic discoveries in crop genetics and agricultural chemicals made in the period just before and immediately after the Second World War. Major new discoveries are now likely, most plant scientists agree, but they will have to come from new findings in photosynthesis, the control of pests, biological nitrogen fixation, and in such basic disciplines as genetics, molecular biology, biochemistry, endocrinology, immunology, and plant physiology. Today, the efforts in these fundamental science fields relating to food and nutrition are very modest.

Recognizing the weaknesses of our present program of food and nutrition research is made easier by looking at the parallel research effort in medicine, another great applied branch of the basic science of biology. With popular public support, consequent generous funding from Congress, and much attention from leading scientists through their academies, the National Institutes of Health have built up some of the most outstanding research enterprises in the world. The programs have been tied into higher education in the universities and are managed with support of and review by peers. Excellent communication has brought these research efforts the respect of the broad scientific community. As a result, if one looks back at the past 20 to 30 years, it is possible to see that American medical biology programs have stimulated and exerted an enormous pull on both the most experienced researchers and the brilliant young scientists considering entry into the field.

The research base in agriculture has been far narrower. It has been much more concerned with lower-level applications than with the more fundamental aspects of food and nutrition. The medical program is now a good example to emulate. A career in agriculture or nutrition research should be made competitive with one in medicine. This goal requires support of research across the whole institutional spectrum, including, for example, post-doctoral fellowships and training programs for professionals in the basic sciences related to food and nutrition. Actually, a strong research program in food and nutrition could be described without exaggeration as basic to the goals of biomedical research. What we eat has a great deal to do with how well we are. Health in the human species, from the beginning embryo in the womb to old age, is related to the nutrients the body receives. Malnutrition is itself an important disease, and good nutrition can be a major factor in preventing, eliminating, or controlling most diseases.

The science of nutrition, which stands astride the disciplines of agriculture and medicine, exemplifies the relationships between these two. Unfortunately, as a consequence of its dual status, nutrition has actually experienced special difficulties in receiving adequate sponsorship. When a field is shared, coordination between the responsible federal agencies (health and agriculture, for example) is hard to arrange. Yet the consumption side of the food problem deserves research on a par with the production side. Such issues as the nature and origin of food preferences, human requirements for nutritional substances, methods of supplementing the nutritional value of crops through the use of additives, and the relation of nutritional status to disease all deserve research emphasis.

The new need for the United States to limit its petroleum imports, when considered relative to agriculture, well illustrates the reasons we should arrange a stronger program of scientific research and technological development in agriculture. While American food production is efficient in output per acre of land or per worker-year, it is inefficient as a converter of energy. The energy required to produce agricultural chemicals, or the energy dissipated by the farm's operating equipment in moving water, earth, and crops and in processing, storing, and delivering the yield—from preparation of the land and planting to finally available food—is several times greater than the amount of energy contained in the food when ready to eat. In lettuce, for example, the efficiency, expressed as a ratio of energy content in lettuce eaten to all the energy required to bring it to our plates, is a mere 1 or 2 percent. This important energy ratio in agriculture can be improved. In the past it has not been important to seek betterment, so very little effort has gone into the needed scientific research and engineering. Minimizing energy is a virgin field for agricultural science and technology to plow and it is becoming one of increasing importance.

The energy required for modern American agriculture is about equally divided between that used to operate farm machinery (for moving and handling of earth, water, and crops) and that required to produce and utilize chemicals (fertilizers, herbicides, pesticides, and preservatives). As to machinery, a tractor has an efficiency between 10 and 20 percent, expressed as the ratio of final work accomplished to the energy initially stored in the consumed fuel. Some fundamental limits would prevent engineers from ever bringing efficiency up close to 100 percent. On the other hand, the design of tractors in the past has not been dominated by a priority to improve this efficiency, so probably gains can be made.

Substantial improvements in the energy ratio of agricultural chemicals probably depend on success in coming to understand much better the basic science of the biochemical processes of plants. Here the need is for advances in genetics and in the means by which plants take up nitrogen, because the largest slice of the energy in chemicals used in agriculture is in the fertilizers that aid nitrogen fixation. What would really help, for example, would be to develop species of grains that, perhaps aided by specially engineered bacteria, would be able to fix nitrogen directly from the surrounding air. Even herbicides and pesticides might be improved by hormones that would work at the genetic level so that far less of these chemicals would be required to achieve effects equal to those now achieved by blanket bombing of the terrain.

Both the machinery and the chemicals now used in agriculture are fossil fuel-powered; the tractors use the fuel and the chemicals are made from it. In principle, chemical rather than agricultural approaches could be used to synthesize carbohydrates, proteins, and other foods. Petroleum would not have to be used, but even if it were, reasonable conversion ratios suggest the penalty might be accept-able even with the shortage of petroleum. This is because the energy content in all the food eaten in the world is only one-tenth of that in the petroleum used in the same time period, and the synthetic production of food by chemical approaches would eliminate the need for petroleum for farm machinery and for fertilizers and other agricultural chemicals for which the efficiency of conversion of energy is low. Both agricultural growing of food and chemical synthesis of it require nitrogen in the form of ammonia. Biomass, rather than petroleum, might become the main source of ammonia. When ammonia is used for the direct synthesis of food substances, no significant loss of material occurs, but when ammonia is applied to the soil as a fertilizer, only a small fraction of the nitrogen is incorporated in the food. All in all, a significant saving in energy and materials could occur if synthetic food production were developed to share the food supply problem with agriculture.

Another entirely different approach to the creation of energy-rich food depends on microorganisms rather than fossil fuels. Bread and wine are two examples of our use of microorganism-induced chemi-cal reactions to convert inedible and non-nutritious raw materials into digestible, preservable, and pleasant tasting food. Further research and development in the use of microorganisms might teach us how to get them to do more of the work now accomplished by the growing of plants in the earth. Plants use solar energy to split water into hydrogen and oxygen; the hydrogen is then used to feed all of the myriad of chemical changes that take place inside the plant. To make this process effective, we now use up a great deal of energy. It is not out of the question for microorganisms to provide the functions required with considerably less energy. If we could produce hydrogen efficiently in other ways (say, by large-scale, safe, efficient nuclear reactors whose generated electricity would decompose water), we could turn that hydrogen over to the microorganisms and let them take it from there.

Finally, we should note that part of the inefficiency of present techniques in agriculture results from the large percentage of waste. When humans first learned to use fire, fueled by waste biological products such as fallen trees, their food supply was greatly increased, not because more food was grown but because so small a fraction of the available vegetation was edible and digestible when eaten raw. Cooking brought a much larger spectrum of the available biomass into the nutrition category, in part because many plants and roots that are toxic when raw become safe after being subjected to high temperatures. Today's farm waste is several times greater in energy content than the food produced, and the grand total of all energy in all biomass the earth produces, including forests, is enormously greater. Of course, converting this biomass to fuel would be one route to support agriculture. It would simply be a matter of creating fuels from biomass, instead of from fossils, to run the machinery and produce the chemicals needed. However, it is also conceivable to find ways to convert this enormous biomass supply directly into palatable and nourishing food. This could be done, in theory, by advances in biochemicals and use of microorganisms.

The method used most effectively in the past to convert biomass to edible food has been by way of food animals. They eat the biomass that humans cannot, and convert it into protein in the form of meat products fit for human consumption. This system can also be improved by further research and development. However, when we change grain into proteins in this manner, the ratio of the energy used to produce the grain to the energy in the protein end product that nourishes human beings is very high, between 10 and 100 to 1. Stating it another way, the efficiency is low, 1 to 10 percent. This poor figure makes it challenging to try doing it much better, skipping the long route through animals and going directly from a combination of sunlight, biomass, microorganisms, and efficiently produced hydrogen to nutritious food.

As another example of research which must look to the federal government for support, let us consider weather prediction. Weather is a crucial factor in agricultural productivity. It is of growing significance for the future as we attempt to grow crops in marginal areas (for example, northern latitudes) and to use genetic strains of plants that provide high yields if the circumstances are ideal but are more vulnerable to small fluctuations in the weather. In any case, it is known that relatively small changes in temperature or rainfall or the arrival time of frost may cause disastrous crop problems. More accurate prediction of the weather, from daily forecasts to six-month projections, would greatly improve crop management and would minimize crop failures.

Through basic meteorology extensions and advancing measurement technology, it appears that more reliable weather forecasting methods are just over the horizon. New remote sensing devices on ocean buoys, balloons, aircraft, and earth-orbiting satellites can greatly improve data gathering. Prediction methods depend critically on the accuracy of information on the current state of the atmosphere and the oceans and on very complex mathematical extrapolations of these data. Now that economical and high-capacity electronic computational power is available, scientists are able to consider much more detailed physical models which should have much higher predictive value. Weather research is not readily commercialized. It will not go ahead through free enterprise alone.

There is a long history of government research in agriculture. Many important facts concerning measurable productivities for land and labor have been recorded and analyzed carefully. The return on investment in research is believed to be in the neighborhood of 50 percent. This figure is remarkably high when compared with the return on investments in most fields of endeavor. If those activities actually sponsored have yielded this high a return, it is suggested that more investing could be very valuable even if not so rewarding as the 50 percent figure indicates. This is the same as saying we are underinvesting in agricultural research. A figure of 25 to 30 percent would beat the average return on investment of the blue chip companies making up the Dow Jones average.

Some, mindful of this evidence of underinvestment, have proposed not only a larger program of agriculture R&D, but a highly centralized Washington bureaucracy to carry it out. This latter proposal could be a mistake. Students of agricultural innovation in the United States believe that a very important factor in past successes has been the decentralization of effort. In each region the farmers and the research laboratories nearby have been in close contact. Scientific research, technology-oriented development, and the operators in the field have been in excellent communication and have cooperated. In any broadened effort, these ties should be preserved.

The third area of endeavor we said must be dominated by the government is the regulating of food activities to protect safety, health, and the environment. The government is already in this role. What else is needed? What needs to be done differently?

We should first note that pure food laws are not the whole answer. For instance, air pollution is already affecting crops. Potato yields are down in Connecticut, spinach is disappearing from vegetable farms near cities, and important grape crops have had to be abandoned in parts of California. We are all aware of the concerns, which are more than a little justified, that mass spraying to achieve improved agricultural yields can influence the health of human beings in adjacent areas and that hazardous matter can transfer from the chemicals used to aid agriculture, by way of farm animals, to the bodies of consumers.

The criticism we have to make, and one that generally applies to all the government's regulatory apparatus and missions, includes more than the usual complaints about bureaucratic delays, lack of adequate competence, and inclination toward over-regulation. These shortcomings of any government regulation effort can be removed, in part, if the government's role is more sensibly defined in the first place. In an ideal world, what would we like to see as the responsibilities of a government unit handling safety, health, and environmental protection in the food and nutrition field?

We would like an entity* that, first of all, has the mission of understanding thoroughly all the hazards of every aspect of the food process from planting to eating. This government unit should be adequately funded and staffed to be able to plan, direct, and operate a program of data gathering and investigation. It should obtain the knowledge it seeks both through its own staff and through contracts with universities, non-profit organizations, and private industry. The highest competence should be searched out and tapped wherever it exists. The agency should lay out the problem to be attacked, pay for the research, monitor the results, and integrate them into a continuing advance of understanding of the area. It should parallel this research with the formulation of standards to which it would suggest all who deal with food and nutrition should adhere. It would then bring its recommendations for bans or modifications of practices to a government decision board. This board, like the proposed nuclear and other decision boards described in earlier chapters, would be appointed by the President and would make tradeoff analyses comparing the benefits and dis-benefits of applying various substances or techniques in the food and nutrition field before reaching its decisions.

As with energy and other areas we have already discussed, the citizenry will have different value judgments about the tradeoffs applying to food and nutrition. There is no single right answer to almost any interaction question that can be posed. Thus, again we have the problem of getting the best decisions made in the interests of society at large when a group (usually from the private sector) wants to

---

*Of course, in keeping with the proposal made earlier in the book, we would prefer it be a unit within the single investigatory agency we called SHEP (Safety, Health and Environmental Protection Agency).

introduce or continue with a food product, additive, preservative, fertilizer, herbicide, pesticide, technique of storage or packaging, or any other component in food and nutrition, while at the same time the government's hazards investigatory agency (SHEP) suggests banning or modifying the practice or innovation to meet its standards. The more competent the government agency that has the mission of understanding the negatives, the sounder will be its recommendations and the more defensible and persuasive will be its advice and conclusions. Its proposals will be less likely to be rejected by the decision board created to settle contested issues. Also, fewer cases of confrontation and disagreement between the proponents of an activity and the government investigators will have to go before the board for final decision. Most often they will be settled "out of court" and the nation will have the best balance of safety, high nutrition, and high production of food to satisfy world demand.

As to the fourth area we listed as a necessary function for government — the allocations of energy, land, and water — we should first note that farm production in many areas of the country is already limited by water shortages. Providing additional water to those with a first interest in farming typically involves substantial controversy because of the interests competing for the available water supply. Thus, the desire of Los Angeles area constituencies for water to support expansion of population and industry there creates continual battles with agricultural areas further north, both sectors wanting to tap the same limited water sources. As to land allocations, when it becomes more evident that agricultural products are in increasing demand in the world market, land use contesting will probably increase because continued urban sprawl and soil erosion will tend to decrease the farmland available.

A clash also is inevitable between the desire to increase agricultural production and gain the benefits of a positive contribution to our trade balance along with higher returns on investment in agriculture, on the one hand, and a national requirement to decrease energy demand, on the other. Government funding and government allocations are particularly complex and sensitive here. Decisions by the government on synthetic liquid fuel development, or conversion of biomass to energy, or conservation of auto fuel, or the balancing of the risks versus the benefits of higher contributions from nuclear energy — all these matters will have a bearing on the availability of energy for agriculture.

Consider just one of many interesting interface examples the government must handle. If the government spends more money on agricultural research, the amount of fossil-fuel energy required for a given amount of agricultural output may be decreased. This would cut down on the requirement for import of oil. In turn, this will improve our balance-of-payments problem. To complete the circle, this then means the urgency to increase agricultural production–in order to increase exports as a means of improving the balance of payments — will be decreased. This reduced urgency will then suggest less emphasis on reducing the energy requirements of agriculture. A happy balance has to be found. The tradeoffs, both economic and political, will have to involve the government in a leadership role, even though the free market should determine the balances to the maximum practical extent. With so much at stake, the individual tradeoffs will produce domestic constituency pressures on the government as will the international implications of what the United States does. The interfaces of government influences with the workings of the free market will be multidimensional. They are certain to be handled on a political basis and, in the end, that is the right way.

There are yet other areas of natural or required government intervention in agriculture in addition to those involving energy, land, and water. There is bound to be increasing opposition by some labor unions to expanded farm mechanization. One instance has surfaced in the tomato harvest dispute between farm workers and the University of California, where automation techniques are under development. This one example, the implementation of the mechanical tomato harvester, a technological innovation that is labor saving, well illustrates the nature of interacting issues which inevitably bring government intervention. The mechanical equipment development was delayed until, through advances on the biological front, a variety of tomato could be created that was suitable for mechanical harvesting. The economics factor, in this instance the cutting of the cost of labor, was adequate to spur technological innovation in the private sector to produce the equipment. However, the technological industry was not the source of the biological advances that produced the new tomato. That, instead, came out of government-sponsored research and development. What the government elects to spend its funds on obviously has a substantial effect on related private-sector investments.

Since land is scarce in Japan, technological developments there focus on increasing the productivity of land use. In the United States, programs have been directed more toward improving the productivity of labor because of the relative abundance of farm land and scarcity of farm labor. However, the United States situation probably is going to change very appreciably. Government policy will have the greatest effect on the speed of technological developments that go to improved land use, decreased energy consumption, or lowered labor requirements.

The trend to bigger farms will probably expand as time goes on. Large farms can be expected to make increasing contributions to agricultural technology advance because they are in the best position to learn while doing. The larger the scale of their operations, the more reasonable it is to expect they will organize the process and apply scientific approaches. It will be sensible for them to accumulate data and, through study and trial and error, improve timing, seeding depth, watering, fertilizing, and other techniques. However, despite the increased volume under control of one farming organization, it probably will still not become the practice for even the largest farms to engage in truly basic research and development. Seeking proprietary techniques and patents does not fit agricultural industry the way it does technological industry engaged in the manufacture of equipment.

Large farm managements will have easier access to capital markets than small farm owners and will be more educated in the application of advancing technology. They will exert an ever increasing lobbying influence on government R&D policies and programs. They will become increasingly sophisticated regarding international markets for their products and the relationship between energy, water, land allocation, and other government-controlled issues and their own return on investment. Agriculture will become more clearly than now a member of the big business fraternity. Government and the agricultural industry probably will develop a relationship similar to that involving other business today—more adversary or contesting than cooperative. For example, the agricultural business lobby could be expected to argue that from a social point of view agriculture deserves high priority in energy use (at least as compared to automobile travel for recreation, or jet travel to all parts of the world for trivial purposes or the manufacture of items that are disposable when they could almost as easily be made reusable).

The worldwide importance of the United States programs in food and nutrition makes it imperative that they be managed exceptionally well. Both the free enterprise and the government sectors will have vital contributions to make in this field, with vast benefits to domestic and international political and social stability. The future responsibilities of the two sectors should be different from those of the past, when neither sector contributed to the fullest. The new roles and missions will take time to develop and mature. When they do, a broader government role in overall systems management in food and nutrition will emerge.

12

the environmental government-private mismatch

A certain river in the U.S. is used as a water supply by several industries whose numerous plants all pour their waste into it, as do the nearby cities. A host of commercial vessels operates busily on the waterway and contributes ample additional contamination. The river, once a source of drinking water, long ago lost its utility for that or for recreation. A portion of its surface caught fire recently and burned a long while because the fire boat's crew directed the river's own water on it. The water that is withdrawn usually is treated before industrial use and, before it is deposited, the waste is treated in conformity with regulatory laws. These treatment actions, the before and the after, are largely uncoordinated. A similar situation exists in another area where a dense and highly industrialized population pollutes an ocean bay.

In the first example, an awakened citizenry decided to attack the problem. A team of experts was assembled and asked to analyze every facet of the river's pollution problem: its causes: its interactions with the needs, potentials, and goals of the area; the options for practical cleanup possibilities. The group of professionals was instructed to gather all necessary data, conduct controlled experiments, conceive and compare alternatives for improvement—all so that the public might see what benefits might result from specific programs and expenditures.

The team included scientists, engineers, businesspeople, economists, academic sociologists, and practical politicians. They began by raising questions. Should the goal of the antipollution effort be to lay a foundation for industrial expansion? What are the inevitable conflicts among the separate interests involved? Can the region's economy support extensive waste control measures? What criteria should be used to define an "unpolluted" river?

They knew people cannot say positively how much they want something unless they know how much they will need to pay to get it. Nevertheless, in a reasonable time, the citizens were shown detailed predicted consequences of continued pollution: deterioration of health, lower economic growth, poorer living standards. Also pictured was what commitment it might take in resources, competitive cost handicaps to industry, bigger bond issues, and higher taxes to incorporate new standards for reducing pollution and realizing a superior environment. Everyone understood that the analytical simulations of the alternatives, ranging from inaction to different positive levels of actions, were only estimates. With people being people and having diverse interests, unresolved differences of opinion surfaced and remained. But some ballpark comparisons could now be pondered. The fundamental proposal received majority support. The region chose to go for a major program to end pollution in the river, and the several cities agreed to set up and jointly support a higher authority empowered to control the river's use.

Meanwhile, at that other location, the one around the common ocean bay, some called for a similar effort, but more were loudly apprehensive about trying it. Several leading politicians and businesspeople (and a few professors) opined that an analytical approach was "maybe good for R&D but not for a practical project . . . ; you can't predict people's reactions to change . . . ; there is a lot more to pollution than technology . . . ; no use wasting time because people will never agree on goals. . . ." One company pointed out that if forced to treat its output waste fully it would suffer sky-high costs. Competitors elsewhere would capture its business, and unemployment would balloon in the area. One mayor said "antipollutionists" were communists. Another claimed it was the other mayors' cities that were ruining the waters. A local philosopher said that technology led to pollution and what the world needed was less technology; this was variously interpreted, but mainly was thought to be *anti-* any kind of antipollution project. A newspaper editorial began, "If we can have good air for an astronaut on the moon, then why not in our cities?" and ended by judging the second problem as impossible to solve because of its political entanglements. And nothing was done.

Although both these examples are fictitious, each has a foundation in reality. The second situation, inaction and social immaturity, is the rule. The first example, going ahead to do what should be done, is a rarity. However, the scientific foundation and engineering methodology for pollution control can be made available. Furthermore, society's needs militate in favor of the use of a scientific, i.e., a logical and objective, attack. The reason the approach is so infrequently used is a missing will, not a lack of means. Because accelerating technology is badly mismatched to lagging social advance, we see about us not only problems from the use of technology, as in pollution from technological industry, but also from the acts of omission, as in the failure to use technology to limit pollution.

Many of America's rivers, bays, and lakes are heavily polluted even though a plethora of federal, state, and local laws are beginning to close in on such situations. The regulatory agencies typically announce maximum allowed effluent flows and then seek to enforce these rules. In an ideal situation, a growing body of helpful science and technology would be available to all concerned. Putting science and technology to use prudently would make possible a degree of control of waterway pollution representing a good balance between eliminating health hazards and safeguarding economic strengths for the industry in the area. Other benefits, tangible and intangible, would emerge also, like improved recreation and pleasanter surroundings. But such appropriate employment of science and technology requires the organizing of a combined government and private effort. Let us see how this actually goes in real life.

Take a large lake into which a number of cities introduce pollutants, from which their industries extract water for a number of purposes, and which also is used for transportation and recreation. It is important to preserve the resource. How can this best be done? Certainly not by reliance on free enterprise alone. There is an obvious role for government in setting standards and policing to see that they are met, but that will accomplish only part of what is needed. Something is missing: overall integrated leadership and management for optimum control, preservation, and use of the resource. This direction would consider the requirements of all the participants and the potential of science and technology to maximize the use. The leadership group would evaluate the benefits versus the costs of alternate policies and approaches. This is mainly a government role, but one the government too seldom performs in the United States. It also involves appropriate contributions by the private sector.

Done suitably, with a mixture of creativity and common sense, the selective depolluting of major lakes, rivers, harbors, and coastal waters stands to yield good returns on the investment in health, quality of life, and economic gain through protection and maintenance of vital natural resources. To achieve practical results, however, additional scientific research and technological work are needed to understand pollution phenomena more deeply, develop superior non-polluting approaches to use of the waters, and design and produce a myriad of specialized systems and equipments. Of course, such technical effort would be less than pertinent or even misguided without attention to all the complex interface problems. Learning quantitatively and qualitatively about chemical and other industrial effluents, power utility outputs, transportation pollutants, and city sewage; developing techniques and assessing the costs of minimizing or limiting or altering these environmental impairments; puzzling out the economic value to the society of cleaner waters (how clean?) — these and numerous other factors and tradeoff issues must be defined and worked out. The decision makers — in the end, the public — must be able to see the options and compare the benefits with the price to be paid.

What company would put money into a private enterpise approach to depolluting a major lake? This would mean investing company funds to settle the many interactions and to develop the new chemical processes, superior purification and fuel burning methods, optimum energy use techniques, and the rest of the required know-how. Many specific and difficult science and engineering advances would be vital for an intelligent, broad changeover to an integrated lower-pollution system. But in what context should the individual technological efforts be chosen, considering the company must seek a return on investment and also satisfy the communities it is hoping to serve? Even if a combine could be created of, say, five large corporations that among them possessed all the technical expertise required, how could they presume to design and install a system that would affect the economy and social makeup of the industrialized communities of millions of citizens distributed around the large lake? Surely, somehow those communities would have to have had the dominant say if the whole project were to emerge successful.

If a private combine were to invest heavily to develop a sound approach and all the backup for its implementation, it is not clear who would be its customer, how the syndicate would appeal to the affected communities to make it all happen, to make a sale. The risk-taking group might offer a proposal to install superior waste disposal equipment for a city, one of many contributing pollution to the lake, but why should that city sacrifice some of its limited funds to reduce pollution of the common waterway unless all other polluters simultaneously agree to do their appropriate part in a balanced program? The polluters would not only have to agree; they also would need to possess the funds to buy their piece of the system.

Suppose the numerous independent segments of the customer community come together by some near-miracle of arrangement making. That is, they decide to buy and stand ready with the funds to contract with the entrepreneurial combine to install the entire developed system. Would not the Justice Department be expected to stop this private conclave of corporations, claiming an illegal joining by them to dominate the market? In truth, if some such partnership sank huge funds into the creation of a depollution system capability, a second such team, equally resourceful as to finance and technology, might be reluctant to enter the field as a competitor.

We are accustomed to the federal government's taking on the funding and leadership for implementation of major resource construction projects, such as large dams, especially when the benefits are seen to accrue to a large section of the country. Taking major rivers, lakes, or ocean bays as examples, it is not overly fanciful to imagine the federal government's setting up an Environmental Project Office (possibly within the Department of Interior or the Department of Health and Welfare) to plan and implement a large-scale depollution activity. We realistically assume the government would not be able, and should not seek, to attract the necessary expertise in-house. Instead, through a competition, the government should select a capable technological firm, or a group of them joined for the purpose, to take on the systems engineering and technical direction of the program. This systems engineering contractor would bring in specialty talent through subcontracts to obtain the data, originate the concepts, and make the analyses to identify alternate approaches. From an array of economic, sociologic, and engineering studies, this contracting apparatus would articulate the basic objectives to be sought in pollution control of the waters.

The cities and industrial operations that utilize the lake should be funded so they can add their specialized inputs. For all these first steps, federal money should be used. The implementations to follow would require larger expenditures and should have more involvement of local governments and industries located on the waterway. The financing for the installations should be shared. As to the cities, their cooperative assignment in what has to be done should come out of negotiations with the Environmental Project Office, with these negotiations guided by the analyses performed by the systems engineering contractor. A city that adds its pollutants to the waterway but is unwilling to share in the funding nevertheless will have to meet certain minimum pollution restrictions. The government's pollution control agency (preferably the SHEP unit proposed earlier) would propose minimum standards in any case. The Environmental Project Office would carry on extensive tradeoff analyses in order to plan projects believed to represent sensible balances between the costs and realizable benefits. Sometimes the Environmental Project Office would clash with SHEP. The Office would propose to do certain things while SHEP, the investigatory agency, would want something else done or would argue that the project, if allowed to go as contemplated, would fail to improve adequately the environmental conditions at issue. If these differences occurred, such matters would be resolved by a decision board appointed by the President, as described in previous chapters.

It would be advantageous to motivate polluters to clean up their effluents by increasing the federal share of required funds as a reward. The cleaner the lake can become because users of the lake deposit less waste, the more the lake will be available and valuable for other purposes. Thus, it would be justified to recognize this in the fund-sharing pattern. While certain quality standards would be enforced, each polluter could be required to pay to deposit its waste above an allowed minimum, the fee being related to the magnitude and makeup of the outpourings. The private industries using the common waterway especially should be given economic motivation to improve their operations. It is not inconceivable that, by offering such direct financial rewards for antipollution advances, all the funding from beginning to end of a waterway depollution project could come from the charges levied on the privileges of delivering effluents into it. Such an approach would stimulate maximum use of the innovative expertise of the private sector to minimize their costs.

Contrast the foregoing proposal with the way we now commonly handle technology developments when they interface with environmental controls. Whatever the project area, some anti-pollution advances usually are made by the private sector as it seeks to meet the government's requirements and at the same time realize a reasonable reward for the financial risks in making investments. But little about the process deliberately enhances the probability that such advances will occur. As to the government, its regulatory agencies are everywhere involved today in setting standards and policing them, but seldom do we see an organized effort that puts it all together. No group is now in charge of considering the interfaces and tradeoffs, to arrive at sensible solutions that meet the combination of social and economic requirements for environmental controls. Lacking this decision-making structure, we miss much of the potential contribution from free enterprise; furthermore, the government activity is usually solely the prevention of negatives. What we have proposed, instead, is government organization for waterway depollution with precisely the needed integration, positive initiative, leadership, and tradeoff decision roles. At the same time the approach motivates the private sector to serve up its most appropriate participation in protecting the environment.

By proposing that the government set up an Environmental Project Office, we are deliberately suggesting a distinct systems management role for the federal government in waterways depollution. Although the Office would discharge its duties largely by using the most expert private-sector contractors, it would have the clear overall management responsibility for cleanup of the nation's waters. Such a management responsibility does not now exist and it is important to distinguish carefully between this suggestion and what the federal government now does through the EPA.

Since 1972, when Congress passed the Clean Water Act, the government has spent around $30 billion to help municipalities clean up their discharges before releasing them to nearby waterways. Congress' plan was for the federal government to provide 75 percent of the costs. The lack of progress, and the problems of overruns, political chicanery, bribes, and unworkable installations indicate that the system of management has been totally inadequate. It is only a slight exaggeration to describe the way it works as follows: The federal government furnishes money, the EPA sets mandatory minimum pollution requirements, and the rest is a combination of local and federal political confusion. Every municipality jumped at the chance to get federal funds. Most gave little attention to understanding their real needs, soon got in over their heads by ordering up plants beyond their requirements, then found the costs escalating and their budgets inadequate for operation and maintenance. The EPA's own estimates are that the program will take from $150 billion to $200 billion and two or more decades to complete. This will make it the most expensive construction project in American history, yet overall management was never structured to be strong enough to accomplish the job. Congress proposed the goal that by 1985 water ought to be sent back into a stream as clean as that taken from it, a "zero discharge" goal that is beyond any management organization.

Since 1972, industry has spent over $25 billion to meet EPA requirements on its effluents into rivers, lakes, and streams. Some 90 percent of industrial companies have kept up to date, satisfying the standards even as their severity has continually increased. Only 30 percent of municipalities have done so with their sewage and runoffs. Good management would appear to justify concentration on the sources whose pollution contribution is the greatest. But this is not happening. The problem of accelerating the cleanup actions of many cities and counties is much more than the present management system can handle.

The proposal we are making here involves three ingredients or departures from the present unsatisfactory approach. First, set up the Environmental Project Office to control the federal funds and to assume unequivocally the overall systems management responsibility for cleaning up the country's waters. Second, make the EPA part of the broad SHEP (Safety, Health, and Environmental Protection) agency which would investigate, understand, and make recommendations regarding water pollution but would not issue mandates or make decisions. Third, create a decision board (like those we have previously described) to resolve differences as required between the SHEP recommendations and the Project Office's initiatives, making all necessary tradeoff, balancing, and value judgements and the final decisions.

Let us turn now to another pollution problem which probably would justify its designation as the champion of the pollution olympics since it probably would win first place in more separate categories than any other pollution issue with which the nation has had to deal. We refer to air pollution from the automobile.

The pollution problem of autos and trucks has a special position because the automotive industry is the country's largest. It employs more people, constitutes a higher fraction of the nation's GNP, uses up more materials, consumes more energy, and is more influential on our way of life, distribution of population, and standard of living than any other industry sector. The horde of automobiles in the United States has been called our second population, the total inventory of vehicles being roughly equal to the human population. The exhaust of automobiles, if not limited, would be a more important contaminant of the air than any other source in most populated areas of the country.

In the United States, the job of one employee out of six depends on the good health of the automobile industry. Half of American rubber production, one-fourth of our glass, and one-fifth of our steel go into cars. It follows that government regulations affecting the design of a car have a potentially enormous effect not only directly, as on unemployment, but indirectly on the overall national economy and America's international competitiveness. What environmental controls the government requires, and the time schedule on which it mandates the meeting of these standards, have a lot to do with the price of cars and the rate at which the public shifts from older cars (presumably lower-MPG, more polluting, and less safe) to newer, more desirable ones. Government actions also dominate the selections by manufacturers as to where to put available funds among competitive requirements for them, including meeting the government regulations, enhancing productivity, and carrying on R&D to improve the product and its manufacture.

Yet, despite the vital importance to the nation of the health of the automotive industry, little evidence exists to suggest that federal regulation of the industry, whether covering emission, safety, or MPG standards, was created with the objective of balancing benefits to the public versus overall social and economic costs. As to safety, the first developments, all clearly useful, were introduced voluntarily by the industry. They became mandatory government requirements later. These include seat belts, safety glass, collision-proof door latches, and the energy-absorbent steering column. No one doubts the value of these safety benefits. Because of innovative and ingenious design, the added increment to the price of a car incorporating these features appears modest in comparison with the safety gains.

The government then stepped in with the dubious airbag, the 5-mile-per-hour bumpers and the interlock of the seat belt with the ignition circuit. That is, the government regulatory bureaucracy simply went ahead inventing additional requirements with nowhere near an adequate amount of testing and tradeoff analyses. Thus, it is not surprising that the public itself rose up and vetoed two of these three items, the airbag and the ignition interlock. As to the new bumpers, they may have saved something in cost of repairs after accidents, but they forced consumers to spend a billion dollars per year to cover the cost of the additional adornment. Hundreds of millions of gallons of extra gasoline have had to be expended annually to handle the added weight. It is not evident that any safety benefit has been attained to balance out these costs.

To this day, we do not know whether the air pollution standards (or the safety and fuel-conserving standards) set by the government are in the right range, everything considered. "Everything" means unemployment, inflation, economic strength, and social and political stability of the nation as well as the specific dis-benefits the mandated controls were intended to correct. When billion-dollar annual losses are sustained by American automotive manufacturers and hundreds of thousands of auto workers are put out of work, this suggests that something more sophisticated is needed than these two points of view which have dominated government decision making: One, the more severe the environmental standards the better. The other, automobile manufacturers, had they been managed with competent attention to air pollution, would have anticipated the requirements, so they have only themselves to blame for their difficulties in meeting government mandated standards. The trouble with this approach, of course, is that the millions of people dependent on the industry suffer from the blows the industry takes, We have to recall that a family whose head is unemployed eats less well and puts off going to the dentist or the doctor, and this constitutes an impairment of the family's health just as does automobile exhaust emissions. Furthermore, high unemployment and the inflating of the price of cars endanger the economic and social stability of the nation.

Had we all been agreed from the beginning that careful balancing of all factors should be the rule for government actions relating to the automotive industry, key tradeoff analyses would have been performed. Admittedly these would have been very difficult to carry through, but such tasks were not even attempted. No government agency was handed the responsibility of performing these interface compromises, pitting potential gains against estimated costs. Instead, with a single set mind, government regulators simply proceeded to go down a list of apparently favorable changes they thought they ought to mandate. When the public rebelled, a retreat or deferring in the steady drive took place, but balancing, tradeoff deliberations still did not occur.

This is criticizing the past. Let us look at the situation ahead. We can illustrate that the problems are not over and will most likely keep piling up, by recalling the matter of diesel cars. How will the government go about deciding what fraction of cars should be allowed to be diesels? The auto industry has asked the EPA for a multiple-year waiver of the emission requirements on diesel cars because the industry does not know how to meet the planned tightening of the nitrogen oxide and particulate standards. If this additional purity is mandated, then the car manufacturers' hopes for diesel cars will be frustrated, whether they be of foreign or American origin. Less than 5 percent of General Motors' 1980 cars will be diesels, but the GM goal is to get diesels up to 15 or 20 percent of total GM cars sold by 1985. Conclusive proof that particulates from diesels cause cancer has not yet been developed and much more research is needed to go from early evidence that diesel exhaust might be highly carcinogenic to solid conclusions. Of course, the EPA could ban diesels altogether, but then the nation would fail to reach the MPG goals that have been set down and that, at the moment, are legal requirements on the automotive manufacturers.

In a similar way, automobile safety standards mandated by the NHTSA (National Highway Traffic Safety Administration), when considered together with the air quality aims of the EPA, add up to a conflict with NHTSA's own announced fuel economy requirements. Generally, increased safety and lowered exhaust emissions add weight and impair efficiency, and both hurt MPG performance. The end result is that automobiles cost a lot more than they otherwise would, so fewer of them are sold. Not only does this increase inflation and unemployment and weaken the industry, but it slows the changeover to the more fuel-efficient cars which is supposed to be a key objective. With sales of American-made automobiles plunging and the auto makers taking heavy losses, the government now is forced, on a purely political basis, to consider relaxing the severity of its regulations. Simply deferring making these regulations more severe, as is now scheduled, could mean billions of dollars to the industry and the consumers and would accelerate the changeover to the superior MPG cars the nation wants.

Whether the manufacturers do or do not deserve special blame, it is a fact that they cannot either finance or physically complete in the desired time a changeover to cars that will readily meet the polluting emissions standards as well as the safety and MPG requirements now being imposed. Even if by innovative breakthroughs the changeover could be accomplished immediately, the cost of the retooling is enormous and the price of the cars would be proportionately high. The comparison, of course, is to foreign cars. They have always been small because of the high gasoline prices in those countries relative to American prices (even the present ones); it is not so much that the foreign car manufacturers brilliantly foresaw the American market's requirements for the 1980s as that their own domestic markets caused them to have something closer to today's required designs in the first place. The end result is that the stricter the American standards in air pollution MPG, safety, and requirements, the more it is guaranteed that in the near term American car manufacturers will lose the American market to foreign manufacturers and that unemployment and economic difficulties will soar in the United States.

We know now that sound solutions about air pollution from automobiles cannot be made in isolation and out of context. Air pollution must be considered as interfacing with an array of other problems that go from narrow but different issues to very broad social–political–

economic ones. Clearly, we need a better way of handling this class of problem than the one we used in the past and are employing now. If we wish to propose improved approaches we can help ourselves by recalling two fundamental considerations that we have had occasion to emphasize so often in this book. We need to arrange that there be a mechanism for comparing benefits against dis-benefits. This is a government function. We also need to assign proper roles to both the government and the private sector to implement policy with actions. In line with these thoughts, we return again to the proposal for a government agency (SHEP) that will investigate negatives but not be involved in tradeoffs and decision making, and a separate, parallel decision board, appointed by the President to manage tradeoff analysis and decision responsibilities.

The automotive industry will certainly make tradeoff analyses to aid in designing and marketing its products. An offering by a manufacturer is a sound one if it meets the government-edicted requirements and is also acceptable to the consumers. SHEP (Safety, Health and Environmental Protection Agency) should investigate every facet conceivably coming under the heading of dis-benefits. It would be hoped that most often the automobiles proposed by the automotive manufacturers would meet the recommendations of SHEP. When they do not, SHEP would oppose the plans and the matter would come before the decision board. The decision board for the automotive field would have to maintain a continuing and substantial analysis activity. If it does its work competently, the chances of its being reversed, whenever an automotive manufacturer or SHEP or a third party should take the issue to the courts, would be small. When occasionally such a reversal should occur, it probably would be because the board will have gone beyond its powers. But most importantly, in the decision board the nation would have a group of individuals chosen to consider the automobile problem in relation to the total United States situation and to make decisions that put it all together.

13

extending
human brainpower

If information is put in the form of digitalized electrical signals, electronics hardware can accomplish remarkable feats in handling that information. Computers, human input keyboards, video display consoles, intelligent terminals, microprocessors, digital communication systems, communications satellites, and wide-band cables, all programmed by software, can now make us smarter at our jobs. The new technology is the result of breakthroughs in the basic understanding of electrical phenomena in matter. A couple of decades ago, functional performance properties of electric circuitry of tremendous complexity required a large room full of wires, plugs, connectors, vacuum tubes, switches, coils, transformers, capacitors, and resistors. Now they can be instilled in a little chip smaller than a fingernail. Information systems based on this technology are able to absorb, store, categorize, process, ponder, transmit, and present information in vastly higher quantities and speeds than ever before attainable, yet their cost and size are radically reduced and they have more reliability and accuracy. Used by each of us as we do our work, this synthetic electronic intelligence will extend our brains and senses.

Information makes the world of human activity go around. It keeps the society's operations moving and controlled. Since most of our time is spent gathering and utilizing information, the new humanity–technology partnership in information handling that is now practical has the potential for increasing enormously the value and hence the productivity of every hour of a person's time. This possibility applies to business and industry, banks, the professions, airlines, hospitals, educational institutions, and even government. Virtually all men, women, and children in the United States could be aided in their work, schooling, traveling, and shopping by the availability of such technology. If innovative electronic information technology were able to improve everyone's output to an extent valued at a mere dollar a day, these savings would justify an investment of a thousand dollars per person for the acquisition of hardware and software. The nationwide installing of the emerging electronic information technology would affect hundreds of millions of us and would call for hundreds of billions of dollars of investment. So we are speaking of a technology with gigantic economic implications.

As with the telephone, automobile, and TV, these electronic information handling advances become highly economical only when the users begin to be numbered in the tens of millions. Regardless of how rapid or slow the pace of the changeover, or what detailed form it may take, the investment in information technology sufficient to bring most of the population into the act almost certainly will be forthcoming because of the eventual high return on the investment. Thus, we should assume we are headed toward a society with great dependence on highly automated information handling. While the magnitude of this development, even when extended over many years, is impressive in financial terms, this is the smallest part of the story. The economic impact is less important than the humanistic aspects of extending and replacing human brainpower by machine "intelligence." The accompanying social–political dilemmas, problems of articulation of choices, difficulties in decision making, and arrangements for sound government–private relationships will be as challenging as any in the whole spectrum of technology–society interactions.

No look at the future is realistic without including the alterations in our mode of life made possible or forced by the entry on this planet of machine-intelligent life. The impact of this technology invasion during the next several decades may turn out to be more profound than advances in the fields of energy, food, medicine, or space. Consider that the industrial revolution of the past two or three centuries, giving us enhancement or replacement of human muscles by mechanical devices, changed the competitive positions of nations, the relationships of material values to other values, and the nature of life of everyone on earth. Everything the human race will next do to influence and utilize the environment depends on brainpower. It does not appear an overstatement, therefore, to suggest that the extension of human intelligence by electronic machines will have an even greater effect upon humanity than the industrial revolution.

If we could wave a wand and turn back time so that we could watch the pyramids being built, we would be disgusted by the improper use of human beings for such a task. To link up thousands of mere body muscles to pull heavy stones, we know now, is inhumane and unsuitable. It is beneath human dignity for a person to be used for nothing but the exerting of pure physical force. For this, machines should be employed. But in addition, human beings are hopelessly inadequate for the job, being both qualitatively and quantitatively ill-matched to its specific requirements.

Most of us do exactly the same thing with our brains most of the time. Sorting data sheets into categories, adding figures, acquiring routine information, filling out forms, moving pieces of paper about, reducing orally obtained knowledge to written or printed form, or retrieving and supplying information from a file — in all such intellectual tasks a human has a trivial capacity compared with easily designed and produced electronic machines. Unaided, each of us can add a couple of single-digit numbers together at a rate of a pair per second. After a moment or two of the rapid addition, most of us become bored and mentally tired and we start making frequent errors. A computer can add two numbers with many digits in a millionth of a second, or a billionth or even much less. Any of us, no matter how good our memories, would be hard-pressed to list in alphabetical order all the people we have ever met. An electronic computer does it cheaply and easily. For such routines, there is no comparison between people and electronics, for speed, capacity, economy, and accuracy.

In principle, we can divide up many of the mental processes we use regularly into a number of successive logical steps. We then can implement these steps one at a time, noting that the human brain requires at least a second for each such incremental move in the thinking process. A machine can make the same move in a billionth or a trillionth of a second. As a high-speed, accurate mass handler of information, an electronic computer is therefore overwhelmingly superior to the human brain. A human thought process typically can be described as a program that combines stored information, a set of logical rules, new input data, and a desire for a specific form of answer. If we understand this program we can not only design an electronic machine to accomplish the thought process faster and cheaper than the human mind can do it, we also can connect our electronic information handling device easily with other individuals and with other electronic machines nearby or great distances away.

Of course, many human thought processes are so complex that, not understanding them, we do not know how to design a machine to perform them. We then flatter ourselves by calling this a higher stratum of thinking and describing it as judgment, creativity, or intuition. But even when we engage in these less well defined thought patterns, our efforts can be supplemented by the tremendous recall, sorting, and comparing capabilities now available through electronic information technology. The human species, by teaming with electronic machine helpers, can step up to a higher intellectual level, exploiting the machines' extended memory, logic, and speed to add variety and more dimensions to the thinking processes. We can allocate to the machine heavy routine tasks for which we are unsuitable. Aided by electronics we can do the rest of our thinking differently and better. The right combination of people and machines embodies a higher plateau of brainpower, both in quantity and quality, than people alone.

These concepts for the expanded, versatile processing of information are being paralleled today by advances in communication engineering. For instance, lasers make it possible now to create highly focused light waves with very precise frequency control. We can communicate by transmitting signals in the form of these light waves along extremely thin glass fibers. Fiber optics systems promise benefits of greater communication capacity at less cost. Twenty-five glass fibers can fit into the same duct that now holds just one copper cable, at one-half the cost, with each fiber able to carry many more signals than the copper line. Similarly, space satellites decrease the costs of long-distance transmission and add more channels and flexibility. Digitalizing the information before it is transmitted improves the signal-to-noise ratio, i.e., the fidelity and accuracy of transmission, and this technique lends itself to the new semiconductor, microminiaturized circuitry that cuts size and cost and adds reliability.

With this introduction to the nature and power of advancing information technology, let us now address the social and political issues this technology generates. Let us see how our lives might change as we try to arrange to derive the benefits of the new technology. What negatives are to be avoided or minimized? We can best discuss this through examples. Take first Electronics Funds Transfer (EFT), the use of electronic signals in money transactions, to replace the present production and flow of billions of pieces of paper a year for money control and accounting. In the future, purchases will be completed, bills paid, orders placed, money deposits made, interest charged or credited, salaries received, taxes extracted, funds exchanged on individuals and companies, bookkeeping accomplished—all through use of electronic signals moved and processed in accordance with set rules. Such a system will be faster, cheaper, more convenient, and more accurate than present paper-shuffling methods. *Electr. Funds Transfer*

The full blossoming of this EFT society will require heavy government involvement. Let us see why. In any widespread utilization of information technology systems, the standardization of equipment and procedures that interface with the human operators will be very important. It would be quite silly to create overlapping electric power distribution systems that can't interconnect, some delivering 60-cycle AC and others supplying 50-cycle, or 90 volts in one block of a city and 125 volts in the next one. Similarly, electronic information systems intended for interconnection throughout the nation must use uniform techniques, not only in the machinery itself but in the electronic language used to represent information. The form of the electronic signals that carry the information, and the design of the many pieces of equipment making up the EFT system, will have to be standardized so that system segments will be compatible and interconnection will be possible. Another reason for a degree of standardization is that many manufacturers must be encouraged to produce the parts of the system; otherwise one company might eventually attain a monopoly, dominating the designs so that only its equipment could make the system work. The government will have to decide, prescribe, and police the standardization.

Beyond this, the government will have to redefine banking and determine how banking business privileges shall be assigned. It will have to decide who will be allowed to participate in and profit from the holding and moving of money in electronic form and under what circumstances, with what rules and controls. It will have to set interest rates on funds stored electronically, create privacy and security regulations, and concoct the rules for responsibilities and liabilities of all entities involved. It will have to describe how the government will interface with private industry and the consumer for control, policing, extraction of taxes, and granting of access to the EFT system. It will need to ensure harmony between the new system and other aspects of the nation's operations. For instance, if all of us were to do our purchasing with a credit card, and cause our credit card grantor to receive our monthly paychecks automatically (crediting the amounts to our credit card accounts), we would need neither checking accounts nor savings accounts. Instead, we would expect to receive interest automatically through some computer's formula if the credit card account averaged positive, and pay interest if we are found overdrawn on the average. The card grantor, whether a bank, oil company, department store, information technology company, new kind of credit utility, or the government (should it decide to take over), would be in the banking business.

In addition, somebody has to decide on rules to say how much credit each of us or the many stores, companies, and corporations involved can be granted since these credit limits relate to fighting inflation—an area in which the government is expected to be the prime leader. One reason why EFT has not already taken over the control of the movement of funds, essentially eliminating most cash, checks, and other papers as vehicles for that motion, is that it is taking time to get the rules of an EFT society properly set up. There is a concern on everyone's part, from those whose money is being handled to those who do the handling, that the technology may be brought into play before the government has established enough regulations to insure privacy, permanency of records, insurance against fraud, liabilities, and losses through errors, and controls against runaway credit granting. The private sector con only go so far in bringing the technology to fruition without the government's completing its side of the task.

Let us shift to another example. The development of information automation technology will bring about profound changes in America's pattern of control of production and distribution. It will affect as well the relationships of industry, government, and the consumer to one another. By bringing all of the information on the related steps of a production operation into close and speedy synchronization, repeated operations could be scheduled with a previously unattainable precision and efficiency so that everything will be at the right place at the right time in the right amount. Suppliers and their customers could be so interconnected in information flow as to eliminate the penalizing peaks, valleys, and bottlenecks caused by materials or parts scarcity or over-buildup of inventory, or lack of machine or labor availability. The mining of fuel, the generation of electric power, the processing of raw materials, the making of parts, and the scheduling of machines and factory labor all need to depend on an interrelated chain of information flow if all steps are to be efficiently intertwined.

The new electronic information systems are already able to provide each individual company with the information needed for more solid control of its operations at higher speed and less cost. Soon the nation's manufacturing entities will discover that their improved efficiency can be enhanced further if their information flow is connected to some sensible extent with the data streams of others on whom they depend. Thus, it will be a natural step for a supplier to obtain partly automated orders through electrical signals from customers, who will in turn obtain some of their schedules from incoming orders that arrive electrically from their customers, and so on. The whole will constitute a national network of production information movement that can eliminate a great deal of presently wasted time and money. Instead of the crude estimates that each unit now makes of what might be expected of it in the period ahead — estimates that often turn out wrong and require costly cushioning and alteration arrangements — a seller's operations will be intimately linked to the on-line, real-time, continuing development of the buyer's demand.

With virtually instantaneous electronic information flow, processing, and display, industry managers will no longer depend on pieces of paper. The flow of materiel and funds will be controlled and recorded by electrical signals that will change continually to reflect the latest status. Automatically generated financial balances will show who owns and owes how much at any given moment. When companies agree to interconnect their information flow to the extent we have described, these agreements will constitute new kinds of contractual relationships and interdependencies. For tax records and for reporting of earnings to the shareholders of corporations, new guiderules will have to be set up to establish the accounting pattern. Who is the legal owner of goods circulating between corporations in response to automatic electrical signals for which neither company is alone responsible and on which each has agreed to be dependent? Who owes whom the interest on funds transferred? When is the transfer legally made? Who is responsible if something breaks down in the information system and there is lost business or lost investment? The government must establish necessary rules to answer these questions. It must regulate and referee.

As the technology advances the consumer will be brought into the loop. Some of the mechanics have already begun to be worked out, as the credit card relentlessly replaces cash and bank checks. We already see electronic input devices in stores (called point-of-sale terminals) that produce signals in response to the entry of a card and the proper pushing of buttons on electronic terminals. These acts start an electronic information flow that ultimately pays the supplier and charges the customer's account.

But there is much more to the potential of information technology as it relates to the consumer's influencing the production and distribution operations of the nation. Imagine that all of us have in our homes a new kind of "two-way" TV set. Instead of telling us, as in present commercials, only about products that have been developed and are for sale, the TV also will disclose what manufacturers are planning to do, subject to the viewers' reactions. With a price reduction incentive for early orders, the two-way set may ask for a commitment to purchase the product described. A car manufacturer might, for example, tell about the model planned for the immediate future and might offer a substantial discount on pre-release orders. Viewers could buy the car by stepping to the TV set and pushing the right buttons to identify themselves and record their purchase commitment. Appropriate cash resources will then leave the purchaser's account and be moved to that of the manufacturer offering the product. Action, or a buying decision, need not be instant or impulsive. The purchaser may want to look at and try samples on display in a nearby store or showroom. But in the course of a typical year in the future, a good many purchases may be ordered ahead of time—food for the freezer and the pantry, toiletries, home furnishings, clothing, automobiles, appliances, vacations, educational courses, and tickets for concerts and sports events.

Information technology developments, already being tried out in some areas, allow the home viewer to request information through TV cable systems. These beginning systems connect the home TV sets, if the correct buttons on the terminal are pressed, to information service companies set up to respond to questions. What we are describing is a later but likely additional development, already technologically feasible, in which the decision to manufacture and distribute for sale is contingent, in part at least, on the consumer response to the proposed product. The understanding may be that the products ordered through the two-way TV will not necessarily be manufactured. Rather, this step, it will be recognized by the purchaser, will be taken only if enough people place orders to justify the manufacturing start-up. Alternatively, the viewer may be told that the price will depend on the number of orders. If the response is weak and the volume to be manufactured is correspondinly low, the price may come out higher but then the purchaser will not be bound to the order. The manufacturer may even indicate a range of prices that relates to the size of the market and the purchaser might elect to punch in a price above which the order is not to count, this being the upper limit of an interest in making a purchase. There obviously are endless detailed possibilities and we don't have to invent them here.

We see that advanced information technology will make possible a national network in which the planning and implementing of production and distribution of products could be in response to consumers' demand, as indicated by their willingness to commit their funds for the products. Such an information system would offer the benefits of sound economic planning combined with a new dimension of free enterprise. It would be a way of making the free market work better and faster. Implementation of manufacture and a detailed financial recording of all transactions would be based on the information introduced into and processed out of the network. The entire chain of activities would commence with an offer to sell on the free market and the consumer's acceptance or rejection of the offer. For instance, a large group of customers, when they commit to an automobile described by means of this system, would in actuality trigger the process of information transfer that, later in the chain, would initiate orders automatically for the right amounts or iron, fuel, and chemicals to produce the steel, paint, rubber, glass, and plastic required. The system's information would be basic to the scheduling of the machines, the workers to make the parts, and the assemblies for the cars they ordered. All this economic planning and production scheduling would be in response to directives from the consumers, not from the government or some big corporation's leadership.

One immediate effect of such a system would be the reduced risk for the entrepreneur and consequently an enlarged opportunity for a would-be producer with attractive and economical ideas. Today an innovator has to have assured capital backing and commit that capital well ahead of the time when manufacturing and marketing efforts are in full swing. The innovator must gamble that the product will be accepted at the price for which it can be manufactured, and distributed at a level of sales volume that can only be guessed at. The only choice is to risk investment, to build capacity and inventory based on this guess. In contrast, when fully assisted by an information technology network tied in to the consumer's TV set, the innovator can put a proposal to prospective customers beforehand in a meaningful way, for the cost of a TV commercial. If the response is favorable, management can go ahead with great confidence because it will actually have serious customers. The needed capital will then be more readily lined up.

With the availability of the new technology, resources will be assigned with higher efficiency and fewer failures, making more resources available for further advances and investments. Good communication improves the quality of everyone's thinking. Someone else views and hears what is offered and comes through with an even better idea and proposal. Competition stimulates innovation, marrying what technology has to offer with what consumers really need and want. Resources can be devoted more than ever then to enhancing creativity and efficiency in every area.

Now, if you are inclined to be optimistic, you can view this widespread use of information technology throughout the society as affording us a kind of enterprise superior to any form and amount of it we have ever enjoyed. What we have described approaches a truly free market, a democratic free enterprise society. What is produced and offered for sale is in direct response to the public and not controlled by a relatively small group of individuals or a government. The efficiency of the process makes for a higher standard of living. Judgments as to what to produce rest on the public's sense of values, around which they make their purchasing decisions, and around which the allocation of national resources then follows. There is added competition and creativity, an improvement in the reward-to-risk ratio for investment, and ease in stretching small investments to larger ones when solid evidence of market acceptance justifies it.

Of course, to afford us these advantages, the information system has to be pervasive and it has truly to be free. It cannot be monopolized by segments of the population who manipulate it for their own ends, or run by a government agency to serve political purposes. If you are inclined to be pessimistic, envision that any future technological society whose production is dominated by a ubiquitous information system network is likely to come under the tight control of the government. Keeping track of and directing every aspect of the operations of our society are made possible for the government by the power of the emerging technology. The U.S.S.R. has been trying to control all facets of its economy for decades with only limited success. The Soviets have lacked neither adequately strong government re- solve nor sufficient subservience in their citizenry. They have merely been trying to get complete control too early, before possessing the technology to make their aims practical. Their available tools did not allow them the necessary information gathering, processing, and dissemination. Without this capability their government was unable effectively to take charge of materials, production, distribution, and the assignment of workers and machines across their large nation with all of its people and resources.

To continue with the pessimistic future possibility, we must note that unfortunately America's advancing information technology is coming during a period when our government is getting bigger and more involved anyway with everything that happens. So it is natural to consider that we may find ourselves with a government-controlled information network, a mammoth automaton. Years into the future, this beast, by then fully matured, will tell us where to work and live, what is to be produced in what quantities and styles, where we should go on vacation, and when to trade in our automobiles. Pickles, housing, and bluejeans will all be scheduled and allocated. In this imagined dark future, the technically feasible system making all this possible will be accepted by the populace on the grounds of greater efficiency and security of employment, fairness, equal opportunity, and the rest. If this nightmare comes true, we will not have to think or decide about what the nation produces, but merely respond to electronic messages. The end result will be that Americans will become anonymous, interchangeable, robot-like cogs in a nation of signals, computers, cables, gears, vehicles, flowing chemicals, and electric power, to a point where it will be difficult to tell the human beings from the machines, our brains from semiconductor microprocessors, and our bodies from the rest of the equipment.

Let us turn back to the optimistic side. Instead of setting us up for a robot society in which we become docile digits with no say in the control of our lives, the same two-way information system could tap the thinking of the public and bring the citizens into prompt impact on national decisions to be made. With a national two-way communications network it would become practical for most of us to spend a few minutes each day receiving information and commenting back into the network on the public issues that interest us. Facts and alternatives could continually be presented to the voting citizens who could just as regularly step to the set, push the buttons to identify themselves as registered voters, and announce their choices of presented alternatives. The technology that makes possible a regimented nation of automatons, can, as far as the technology is concerned, be employed in this entirely different way.

Since resources are always limited and choices have to be made, why not determine national priorities, at least in part, by direct voting? Why not ask people whether they want to work longer hours and have more material goods and services or shorter hours and less? Are they desirous of the privilege of driving their cars more for pleasure in the summer even if that means wearing sweaters in their homes during the winter? Do they prefer four long workdays to save gasoline going to and from work, or five shorter days and less gas for discretionary driving? Are they so concerned about the possibility of a Soviet military strike that they want more of our GNP to go into military preparedness, or are they willing to take their chances on this and enjoy a higher living standard? We already poll today to determine how the public is reacting to issues, and vote on propositions and referenda by laborious methods. Why not poll and vote more directly and thoroughly, especially if the voting is preceded by superior dissemination of facts and identified alternatives?

We might or might not want any amount of this kind of "instant democracy." The technology's availability does not force it on us. We can choose whether to employ it, and precisely the extent. At any rate, if information technology advances are going to alter the way our democracy works, we can be the deciders. Changes should occur because of us, not to us. But this can be only if we are aware of the power, the positives, and the negatives of the technology as we prepare to apply it.

TV, a technological development, we note, already has altered the process of selection and election of a President of the United States. How a candidate looks and acts on TV, how he responds there to questions, has a great deal to do with the way that candidate fares in an election. But the influence of TV on our democratic process is accidental; it merely happened to us. We did not parallel TV's technological development with insight, study, and intelligent participation by the citizens to see how this technological tool should best be employed in the political arenas. Present TV, however, is only a small example of what is possible with coming information technology. Looking ahead, the advent of widespread two-way information flow could change us from relatively apathetic citizens — many of whom rate our real impact so low we don't bother to vote — to avidly interested ones, daily participating in the democratic process. This is hardly something to ignore in thinking about the impact of advancing technology on society.

It is possible that the extension of human brainpower by electronic information technology will be the field of science and technology with the greatest overall bearing on the society for the next several decades. Let us consider the proper roles for free enterprise and government in this potent area. We want to use both sectors so as to attain the potential benefits, but we also must minimize unwanted dislocations and undesirable, forced changes in the structure of our society that could arise from the speed and depth of this technological advance.

Since information technology can increase efficiency, decrease costs, and increase return on investment, there is ample motivation in the private sector for its being applied at a rapid rate. This is already happening. Airlines automating reservations handling, accounting departments getting out bills for business firms, factories scheduling production flow and optimizing machine usage, stockbrokers processing bidding data and trades, department stores controlling inventory and purchasing information at the point of sale—these are typical applications of present information technology. No action by the government, or the writers and readers of books of this kind, appears to be needed to match up the potential of the new technology with money-saving systems in private industry.

To make the fullest use of information technology in the business, industrial, professional, and educational efforts of the private sector requires two kinds of competence. One is in understanding the operation for which information handling is being automated and improved. Here the experts are those running the operations: banks, manufacturing companies, department stores, airlines, etc. If these specialized managers do not know their business in detail, no one else does. The other kind of competence is in the new technology itself. Here the expertise is in the technological industry specializing in information handling, those who design the computer systems and their components, digital communications networks, intelligent terminals, input/output devices for interface with human beings, and memory systems. These two kinds of private-sector experts will find each other, shake hands without the government's making the introductions, and come up with the creative ideas and the practical hardware and software embodiments to permit going after and realizing the rewards.

We certainly need nothing from the government here by way of stimulation. But we do need to have the government ably and adequately engaged in a group of functions that only the government can provide. We also need to be sure the government does not wander into areas of endeavor that can only encumber and delay information technology advance. What are the essential government tasks?

One of the many roles the government must handle is that of sponsoring basic research in the universities to back up information technology advance. Again, we speak here of research so long-range and fundamental that it cannot readily be associated with specific product endeavors. All we have said earlier about the government's arranging for generous, flexible support of the universities with a minimum of administrative red tape can be repeated. Since information technology is a field that will see great commercial expansion, it is sensible to imagine that the government-sponsored research should include an extraordinary effort to be sure the research results are broadcast to the industry at large. Some of the government research budget should finance the universities so they can stage frequent conferences and publish the results widely. Very large corporations, some actually involved in joint programs with universities, will be certain to follow up the basic university research potentially influential to their future products. But small businesses, vital to the advance of information technology, cannot afford to staff up for detailed, continuing interface with college research programs.

We should mention that government R&D in information technology on behalf of the government's missions in defense, space, biomedicine, oceans, weather, energy, and other areas will have a great bearing on the strength of the nation in information technology. The government's total expenditures to advance information technology are about equal to the total of the private sector's R&D in this field. Although the objectives of the government in these special missions are quite different from those of private technological industry, considerable relationship exists between advances in the government's fields and future applications in private business and industry. This interface should be recognized as important if the goal is our overall national strength in this outstandingly effective technology.

Returning to the commercial development of information technology we next note the importance to the entire world of the United States market. The rest of the world has an interest in and a need for information technology comparable with our own. The magnitude of overseas operations that can benefit from the productivity and other improvements stemming from innovative application of information technology is roughly equal to that of America. However, some differences favor the American market as an arena of effort by technological industry, as compared, say, to Western Europe and Japan. We are the largest unified economic entity. Design, production, marketing, distribution, and maintenance in the field can be handled on a national basis with great efficiencies by companies pursuing the American market. By comparison the European Common Market is only partially common and substantially fragmented. Each country has information technology activities in which the government has the objective of strengthening the local or national industry dealing in information technology with respect to other nations' competitive companies.

For example, in France the government took the initiative to cause the merging of a number of computer companies to create a single entity more able to compete in the French market against IBM. To achieve the results intended, the French government arranged that most of its own internal needs for information handling be filled by that French computer company. Japan is a highly integrated single economy and its government works closely with its computer industry and Japanese financial sources. It adjusts its rules regarding competition to ensure the health of the home-based industry. However, the total Japanese market is much smaller than ours. No world corporation providing a full line of computers, a prime example of information technology hardware, can expect to compete well anywhere in the world unless it also enters the United States market.

This means our policies with regard to government business (the government as a customer, for its in-house information technology systems and military activities) must be handled with sensitivity to world competition. It cannot be said that the United States government has recognized the problem fully and handled it well. For instance, aside from its roles as principal customer and basic research sponsor, the government has the task of influencing what happens in information technology in the nation so as to prevent unlawful monopolies. Present antitrust legislation and the interpretation of it are, however, not adequate to match up well with the enormous potential for economic strength for the nation in information technology. Thus, the government is bogged down with cases that have been going on for years against AT&T and IBM. While the government studies these two situations, readying itself to go before the courts to claim antitrust violation by AT&T for some reasons and IBM for others, the pertinent technology and the markets change. With these changes the competitive industry's makeup alters and the case preparations are out of date before they can be presented in court. Unless government policies and tactics are altered, it would appear that the AT&T and IBM cases may become permanent phenomena in the United States courts, approaching but never reaching resolution. Nothing about the way the government is conducting these antitrust cases suggests a recognition that the rest of the world exists.

Aside from the government's interest in breaking up AT&T for antitrust reasons, the size and nature of that company create some interesting challenges to government as it ponders the direct roles of both the government and AT&T in the scheme of things relating to information technology. This is especially true as AT&T's traditional role in supplying telephone service is becoming redefined because of the expansion of telecommunications technology. AT&T is America's largest corporation, with assets well over $100 billion. It has one million employees and three million stockholders. What happens to AT&T constitutes a significant part of what happens to America. But more especially, modern technological advances related to AT&T's business make it difficult now to separate or put boundaries around its mission. For example, should AT&T be ruled out of the data processing field? Then what about the gray areas between what might be defined as telecommunications, on the one hand, and data processing on the other. Information handling by electronic means includes telephonic communications, data processing, and video transmissions within and outside of homes, businesses, schools, hospitals, and government. This constitutes a market worth over a hundred billion dollars a year, estimated to reach $300 billion in another decade. The boundary issues of the supposedly separate telephone and computer fields are now so complex technologically that even the specialists have difficulty comprehending them. The breadth of the advancing technology is such that Congress has had to consider legislation to overhaul virtually all the nation's communication laws. In the process, AT&T, previously restricted (by a consent decree in 1956), may become free to use its resources to enter new businesses. This could put AT&T in partial direct competition with IBM and other computer companies. Advancing technology has simply blurred the distinctions among the various, previously well defined, and separate areas of information technology.

The latest communications technology sends the human voice over phone wires or by satellites or through microwave links in the same kind of coded digital bits that a computer uses in its calculations and for storage of information. People and computers communicate with each other. Furthermore, communications networks, even telephone networks, are now becoming intelligent computation systems programmed to achieve optimum services automatically in response to a customer's needs. Already the government has had to involve itself in deregulation and reform as various competitors have challenged the regulated monopoly which AT&T constitutes. For example, in 1968 the FCC permitted independent equipment manufacturers to connect their own telephones and data terminals into AT&T's network. Then in 1969 it allowed specialized carriers to offer their own long-distance service through links set up independently but connected with Bell's local systems. Now both the House and the Senate are working on their own new "Telecommunications Act of 1980." This Act would completely deregulate the manufacturing and sale of all kinds of terminal equipment and would merely require that the specialized carriers pay an access fee for hookups into AT&T's local systems.

Fundamental to all of this is the permanence of government involvement. The government must continue to regulate and assign monopoly privileges to AT&T and other companies to provide certain classes of services as exemplified by the conventional telephone in homes and offices and the cable TV systems of an increasing fraction of all homes and buildings. At the same time, it needs to allow the fullest competition wherever possible, in every aspect of information technology. All of this must be done despite the overlap between areas of regulation and those of free competition. The quality of the government's work will have a great deal to do with the degree and success of technological advance in the United States in information technology.

The government must be the allocator for the limited resources of the radio frequency spectrum. Room is available only for a certain number of broadcast channels of transmitted information and it is no easy task to decide who among competing groups should be granted the privileges of monopolizing parts of the spectrum. Unfortunately, only by such exclusivity in assignment can we have effective transmission of information through the radio spectrum.* The allocations must be in the public interest; exactly what that means has always been difficult to judge. Now, with the avalanche of new dimensions of information transmission and handling opened up by advanced information electronics, the government must rise to a much higher level of competence in analyzing still more options, tradeoffs, and relevant economic–social–political criteria.

Traditionally, technology has expanded so that as the worldwide demand for use of the radio spectrum grew so did the ways in which more and more of the spectrum could be exploited by going to higher and higher frequencies. Also, as time passed, we learned how signals could be controlled more precisely and separated reliably so as to use more efficiently all of the available spectrum. To show how the problem is now becoming more difficult—on an international basis—consider the advent of the communications satellite. Satellites now make possible long-distance transportation of information with lower costs, higher area coverage, and greater information handling capacity. They bring onto the stage many new economically and socially attractive capabilities. The satellites communicate with ground stations which act as the beginnings and ends of networks involving cable and radio systems working ground-to-ground. Cable networks into cities, and into commercial buildings and residences within the cities, have to be handled by further monopoly assignments.

---

*New technological advances may make violating this rule practical. But this will add to, rather than reduce, the government's difficulties in assigning licenses to broadcast.

As with the present telephone systems, the privileges have to be assigned by specific geographical areas.

A communications satellite must be stationary; that is to say, it must take 24 hours to rotate around the earth in its orbit so that, relative to the earth below, it will remain in the same position as the earth turns. The orbit that makes this possible is one that is above the equator at an altitude of about 22,300 miles. First of all, a new spatial problem arises, that of spacing communications satellites of various nations so that they will not produce undesirable interference. Next, no one country's satellite can be up at that distance without its producing radio emanations that could be picked up and become disturbing to virtually one-half the earth's surface, no matter how well the radiation is beamed so as to minimize this interference. Some communications satellites, of course, have the specific purpose of broadcasting their signals over a substantial area while others are designed to pinpoint their targeted areas. Technologically, then, the matter becomes very complex because the allocations are involved not only with space separation and radio spectrum allocations, but with specific rules governing the quality of the focusing of the radiation from individual satellites to individual points on earth.

It would be difficult to understand the phenomena and to set rules, taking proper account of the economics and technology, even if the world constituted one nation and one government made the decisions. On a real-world international basis the problem is much tougher. The speed and economic efficiency with which advanced information technology incorporating satellite systems is employed is thus greatly dependent on the pattern of government and private industry cooperations, on the one hand, and the political strength and diplomatic skill of the various governments involved, on the other. Periodically, a world radio conference is convened to settle the dividing up of the frequency spectrum (and now also to apportion orbits and positions for communications satellites). Current conferences have over 150 nations. The lesser developed countries (LDCs) of Asia, Latin America, and Africa are in the majority. They are determined to challenge the dominance of the industrialized nations, which enter such negotiations with the concept that they are entitled to exclusive use of most of the valuable and limited spectrum because they were the first users. The LDCs do not agree.

In these negotiations, politics, economics, and technology all intertwine, with each nation, young or old, wanting advantageous and unfettered domestic communications networks. In addition to telephony and television, many other requirements exist, ranging from data processing to earth resources mapping and weather and navigation signals. Most of these involve satellites. Those nations that do not as yet have the resources and applications jusfifying complex, broad systems nevertheless want to reserve places in the sky and the spectrum for planned or hoped-for future projects. The more technology advances, the more it becomes possible to approach the miracle of accommodating the needs and the wishes of all. This stretching of potential is accomplished with ever larger satellites, more cable systems to replace radio, and more complicated, albeit more expensive, systems that so code the information that it allows separate transmissions on the same frequency with good fidelity and accuracy of signals and a minimum of interference.

The success of the United States government representatives in these negotiations goes to their having the ability to do a full-systems, interdisciplinary job. They need expertise in existing and future technology and in the relationship of the technology to matters of American economic strength, growth, and national security. For the United States to maintain leadership in information technology it is in the interest of the private sector to do what it can to cooperate and help ensure strength for accomplishing the integration and negotiation tasks in the government, where these tasks belong.

The government has been slow and cautious in allowing citizens access to the benefits of the broadened technology. For example, a sports, music, entertainment, or educational program originating anywhere in the country now can be moved by cable to a nearby large ground station antenna, beamed up to a space satellite and back down to a distant ground station for distribution to its environs via its local cable television system. In this way, with the satellite making distant signals economical to transmit and deliver, local cables can carry dozens of channels, giving their viewers the possibility of choosing many program possibilities. In principle, thousands of programs, individually originating in any part of the country, can be sent up to a satellite then redirected down to small rooftop antennas on a hundred million private homes. Programs can be sent up to one satellite for a lateral pass of the electronic signals to a second distant satellite that, in turn, can rain the signals down on a land area halfway around the world, to be received there by a small antenna. For years the government has restricted this kind of broad program distribution. The public interest would be best served if the government minimized regulation and allowed the maximum of competition among local broadcasters, national networks of video broadcasters, and numerous forms of cable-delivered programs reaching offices, homes, factories, and schools.

The Federal Communications Commission (FCC), which allocates portions of the frequency spectrum for radio broadcast use, has been only slowly lifting unnecessary restrictions on the diffusion of television programs by way of cable, but it is now adjusting its criteria for licensing broadcasters to account for the impact of the advancing technology just described, which so greatly broadens the marketplace for competitive attention of audiences. Under the "Fairness Doctrine," it used to be very important that the FCC enforce such concepts as "equal time" and otherwise ensure that radio and television broadcasters provide reasonable editorial balance in their programming. A broadcasting company equipped with a monopoly for use of an important portion of the spectrum had in part a captive audience that it could choose to misuse (although with great economic and legal risk because its license could be revoked). Now, with the introduction of the UHF channels for broadcast TV and the almost unlimited potential for the audiences to select a wide range of programs opened up by cable and satellite transmission, not to mention such other competition in home entertainment as video discs, suddenly public interests suggest heavier reliance on the marketplace. As with newspapers and magazines, each television viewer will be able in the future to choose from a large variety of specialized programs. Satisfying an equally wide range of consumers' interests will put competitive pressure on broadcasters, which should remove much of the concern that a broadcaster might use a monopoly to propagandize for special causes in an unfair way. No broadcaster in the future will command a captive audience because there will be so many other programs to shift to.

Let us turn now to the important, new two-way transmission by cable. This has now become practical, greatly increasing the possibility of attaining superior education and merchandising as described earlier in the chapter. Here government decision making is even more difficult and ponderous and society may have to wait many years after economical, reliable technology arrives before it can enjoy the advantages. Aside from their use in homes to make possible political polling and direct purchasing and ordering by the audience, two-way TV systems allow viewers to request information. In principle, such two-way information-request systems involve automatic switching of the requester to information sources of various kinds. Somehow the government has to decide to whom it will grant the privilege to provide information on request. As a minimum, some kind of quality control over the service would seem to be necessary. Only the government can be in the control position here, so another interface of government with the private sector has to be worked out. The organizational steps, if long delayed, will hold up the availability of the technology and the denial of its benefits to the public.

There is even a competitive interface between the telephone line coming into each home and the two-way TV cable just discussed. With the advent of low-priced, reliable, and very versatile electronic computers for the home, a host of additional services can now be performed for the homeowner, perhaps using either of the two lines coming into the home — the TV cable or the telephone cable. For example, wiring in the home can permit the telephone circuit or the two-way TV cable, when it is not otherwise being used, to watch and report on home operating functions: reading meters, warning on fires and burglars and sending out alarms, paying utility bills, and receiving status reports on bank accounts, debts, house equipment maintenance, and reorder times on various services. Here again, only the government can do the refereeing and allocation of roles and missions between the two lines entering a home or commercial area.

In information technology a number of government agencies are involved, each with relatively narrow missions. For instance, commercial information technology systems that shift ownership of large funds among financial institutions have a need for high privacy in information handling. Such matters as credit ratings on individuals and the data that create these ratings also must be regarded as confidential. Certain agencies of the government are concerned about invasions of individual privacy. Meanwhile, other government units are supposed to arrange total public access to all information. Some information, government and private, should be available to the public at large, in the public interest. Other information should be reserved for people with proper access, in order to protect privacy. This means computerized information systems need to be equipped with appropriate codes and keys.

These semiopposing privacy and disclosure requirements present further difficulties for the government. Consider the following example. The National Security Agency of the government deals with highly secure military and intelligence information and designs codes and keys for control of flow and access to this closely guarded information. A few years ago private inventors filed for a patent on a scheme claimed to be an advanced, superior, and lower-cost means to protect the privacy of telephone conversations and other information flow. The device is a scrambler, harder to unscramble than existing approaches, it was said. The inventors proposed to sell their devices to anyone wanting to deny others the possibility of eavesdropping on their conversations or data transmittals. The patent application machinery ground to a halt when the National Security Agency was granted a secrecy order on the patent application, claiming that dissemination of the device would hamper national security. The secrecy order even prohibited the inventors from discussing their ideas, let alone selling equipment based on them. This action by the government threatened to preclude the use of the new technology to meet civilian privacy requirements. The conflict is being resolved by the courts, which seem so far to have ruled against the secrecy bans and for allowing the privacy inventions to be employed for general public use.

As mentioned earlier, EFT (electronics funds transfer) requires that the government redefine banking. For very good reasons it has been accepted that the government should regulate the banking industry. EFT and the creation of nationwide credit-handling activities together change the meaning of loans. An unpaid balance on a credit card is the same as a loan. A broadening of the regulation of interest rates is required in order to cover the new ways in which the borrowing of funds, or its equivalent, can be arranged. All these considerations need to be examined by entities in the government that deal with banking as distinct from the radio spectrum licensing, antitrust, privacy, export–import, military, and university research aspects of government involvements with information technology.

One final example, the use of electronic techniques to replace ordinary mail service by the Postal Service, will suffice to illustrate the diversity of government–private relations problems that need to be solved if the advance of information technology is to be fully exploited by America. Here again we have a situation in which more than one agency of the federal government seems to have responsibilities to make decisions. In a continuing dispute with the United States Postal Service over electronic mail, the FCC recently denied the Postal Service permission to send messages across the Altantic. The Service wanted to start an international electronic mail service with letters or messages sent by wire or other electronic means from one office to another and then delivered by hand. Mailgrams are an example of this service. The FCC stepped in, claiming jurisdiction over such service, while the Postal Service and the Postal Rate Commission thought that they had the jurisdiction. The FCC took the position that the Postal Service was intruding into electronic communications, an area up until now handled entirely by private corporations. It argued that for the Postal Service to get into this field would wipe out competition and stunt the growth of electronic communications in the private sector. At any rate, it was not about to agree to the Postal

Service's idea that it should be the exclusive provider of such services and should be granted the sole privilege of using satellites for it.

More recently, the Postal Rate Commission told the Postal Service that it could go ahead with an experimental operation through 1983, provided that the electronic transmission part of the service is handled exclusively by private companies. Private companies, according to the plan, would be allowed to receive and transmit electronic messages to some 25 post offices specially equipped to handle the new service. The Postal Service officials indicated their disappointment that the FCC did not grant their request for complete control.

The Board of Governors of the Postal Service is insisting upon managerial control and ultimate full responsibility for all electronic mail transmission. The Board has worried that the Postal Service could easily lose 80 percent of its mail volume to electronic systems over the next few years and is not about to see all that business lost to private electronics concerns. On the other side, private groups will probably continue to fight for a major role in shaping the development of electronic mail. They argue there is a need for speeding up the delivery, especially of mass mailings like advertisements and bills from businesses to their customers, a type of mail service that has deteriorated in recent years. The industry also is concerned that the Board's decision is merely the first step in an attempt to reclassify all electronic messages as first-class mail, over which the Postal Service has a legislative monopoly, at least by some interpretations of the legislation.

The electronic mail controversy will doubtless turn out to be a typical example, not an exception, in the array of government–private relationships and assignments of roles and missions that the government must lead in working out. Failure to get these assignments of missions set soundly and in a timely fashion will greatly limit the speed with which valuable electronic technology can be made available. For every year in which electronic mail is delayed by inadequate arrangement making, even though the technology is ready, the unrealized savings will certainly be in the hundreds of millions of dollars in direct cost differences. Still further indirect savings will be missed in not having the superior, faster, more reliable electronic mail instead of the present increasingly slow and unreliable methods.

Should Congress create a new government agency to take on the leadership of any and all aspects of information technology—a National Information Systems Agency? Most information technology functions involve key considerations that are outside the information technology area per se. These issues require a different kind of coordination. For example, controlling transfer to other countries of information technology would seem to belong in that agency of government that is in charge of all technology export in any field: energy, military, agriculture, ocean science, or whatever. The Federal Reserve Bank needs to be involved in electronics funds transfer to redefine banking. A single agency could hardly provide clear, concentrated administration of all the government's duties related to information technology. However, this does not mean we have to settle for the present wide dispersal of functions. We could set up a government unit within the executive branch* whose mission would include much, but not all, of the leadership, licensing, and sponsorship aspects of information technology. First, we could house in that one agency those government functions that can be separated away from other activities without great dislocations. Then we could give the new agency the further responsibility of being a recommender of integrated policy even though the agency could not be granted the power to control all government decisions related to this technology. It could analyze all problems involving information technology, study options, and come out with its conclusions and advice. In some instances, its charter would enable it to act directly to implement its proposals. In other instances, while limited to being only a recommender, it still could be a valuable influence on other government agencies whose purposes are narrower.

If the government performs well in providing integrated leadership in those functions it absolutely must handle in information technology, the private sector producing the technological advances also will do a superior job. Those contemplating risk investments will see a more rational, stable, continuing, long-range information technology advance. The free market will then work to optimize investments. This field is of such importance to the United States that the leverage and payoff of outstanding performance of the appropriate missions of the government and the public sector would yield highly satisfying rewards.

*The present FCC might serve as the initial building block of the proposed broader agency.

14

the international dimension

If the United States were the only country on earth, we would still have our hands full harmonizing technology's advance with society. If neither international rivalry nor independent action of other nations could ever bother or affect us, the problem of getting our democracy's goals squared away would not go away. The voters would still display a disconcerting spread of judgments as to risks they are willing to accept to gain various benefits. We would still have to pit more output against cleaner air, lower production costs against greater safety on the job. We would have to work out the relationship between government and free enterprise as we seek to employ well advancing science and technology. We would want to preserve individual freedom, enjoy the motivations and flexibility of free enterprise, and keep the government from running our lives. At the same time, we would continue to insist the government set rules and enforce them, and perform any other needed tasks and functions free enterprise cannot. We would still be blessed with the challenge of organizing ourselves to handle a myriad of issues, to make the hybrid society work.

But the world is full of other nations and they are heavily involved with us. They are autonomous in most ways, yet we and they savor or suffer numerous interdependencies that cannot be ignored. Every country, communist or non-communist, developed or underdeveloped, now believes a close correlation exists between its status in science and technology and its chances of meeting its national goals. Technological resources are perceived as essential for the fulfillment of the material requirements of its people, its economic competitiveness in international trade, and its military security. Even nations that have huge petroleum reserves and could choose to live off the revenues from sale of this raw material to other nations have looked ahead to the period when those stocks will dwindle. Concluding they then must be technological nations, they are spending feverishly to elevate themselves to this category as rapidly as possible. Some nations attempt to expand their science and technology base almost entirely by government initiative; for them, a free enterprise area is either non-existent or secondary. In other countries both the government and the private sector press for the acquisition of higher technology and the development of products to sell to the world at large.

Just as an American company would be unlikely to develop a new computer with the idea of selling it only on the Pacific Coast, so increasingly do technological corporations in all the non-communist world, the United States included, now find it necessary to consider a wider market than the single nation in which that business entity headquarters. If a private technological company fails to go after the world market, it gives its competitors a volume edge. It also passes up product application ideas and sources of knowledge of future technology and markets. Many areas of the world contain customers for whom the product of the manufacturer, if it is successful in the country of origin, may be equally suitable. These are sales given away to the competition if operating internationally is disavowed. Furthermore, for many technological products, numerous applications exist. These applications change with time as the system of which the product is a part is altered by further scientific discovery and technological advance as well as because of changes in the market and the society. These advances and changes can originate in innumerable places in the world. A corporation that deals with a large array of customers in a variety of applications in different countries will be likely to see more advantageous possibilities to exploit.

By proper coordination and integration of activities on an international basis a company producing technological products can often lower its costs and attain either increased returns or lower prices and broadened markets for its technological products. Whether the product is a computer or a four-cylinder automobile engine, there are advantages to choosing the optimum locations for manufacturing facilities and carefully allocating the production to the geographically arranged sources. Instead of an IBM or a General Motors keeping its overseas operations essentially independent of one another, for example, it often can do better by purposely making them somewhat dependent upon each other. The assembly of a product to be sold in Country A may best take place in that country with parts coming from Countries B, C, and D. In this way, the advantages of each individual country—technological strength or labor availability or political stability or volume of local market—can all be integrated into a plan that maximizes performance parameters on a global scale, benefiting not only the company's financial performance but its employees and customers everywhere.

Economic interactions among the nations of the world now involve a contest between two approaches, free trade and isolationism. In a totally free trade world (a theoretical extreme) everything would move freely across national boundaries: raw materials, investment funds, cash flow from profitable operations, technology, management know-how, labor, and finished products. Each nation would make available to the others what it can best supply in a competitive market. One nation might have a plentiful source of raw materials; another, cheaper labor; a third, greater know-how. These different attributes of the many nations would vary with the area of endeavor or the nature of the resources involved. The net result would be a free market optimization in which all nations would enjoy a higher standard of living than otherwise because they would fill their needs at the lowest prices and could sell what they produce everywhere without restrictions.

The other approach, essentially the opposite of the free motion of money and goods just described, is an international economic relationship based on national barriers. Social–political–economic problems in each nation cause its people to demand solutions from their governments. This inevitably causes each government to seek economic advantage over other nations by control of their international trade and the movement of money, physical materiel, and technological know-how. Thus, to protect a local industry from foreign competition, which means to subsidize it, competitive imports are discouraged by tariff and other barriers. Similarly, each nation tries to handle its inflation and unemployment, the exchange relationship of its currency to that of others, its money transfers, and its balance of exports and imports by numerous measures and countermeasures vis-a-vis the actions of other nations. Seeking short-term objectives in response to political pressures, governments interfere with free world trade and incur both short- and long-term negatives as a consequence, even if they succeed in attaining some near-term pluses.

At the present time, the second approach, isolationism and protectionism, may be somewhat in the lead. But the contest is close and may continue for a century. At any rate, the world is not now progressing with alacrity and certainty toward a totally free market environment. However, despite the domestic situations of nations and the resulting government actions limiting trade and money transfers, insistent pressure exists for a substantial degree of unfettered exchange. This is because free trade is a way to raise world economic strength and create growth. Free trade gives each nation the benefits of the resources, know-how, and overall output of other nations while it is maximizing the value of its own assets and talents. Because a high economic payback results from free trade, this potential exerts a powerful force against over-commitments by governments to economic isolationism. In addition, each nation is anxious to export even as it curbs imports, all the while knowing that, since an exported item by one nation is an import to another, one goal is directly in opposition to the other.

The world thus combines open trade and protectionism. It consists of a family of semiautonomous entities each of whose use of science and technology affects how all the others employ these tools. The private and government sectors of all the nations, universally seeking security and economic strength, are always engaging in an enormously complicated set of bouts and deals. American domestic goals, even if we could set them clearly, are bound to be partly circumvented, frustrated, or bent by the policies and implementations of others not under our control. The environment of the world — physical, economic, political, military, and social — is determined only partly by our own situations, perceptions, and decisions. Thus, in the United States, as we assign roles and missions to government and free enterprise in various project areas, our choices are limited by, and our organizational arrangements inevitably bound up with, goings on elsewhere in the world.

Let us look at the good side of this first. Take the matter of providing for the health, happiness, security, and economic welfare of the American people. As we earlier estimated, we can expect more than half of the important advances in science and technology to come from the work of scientists and engineers in other parts of the world. If America is virtually an open market for the products, services, and know-how created in other nations, we can benefit from those developments as they become available, unless we have erected barriers against them or lack the wherewithal for their purchase. If all were well with the world — economically, socially, and politically, with no military threats — we should all welcome a free world for imports and exports. Neither we nor the other nations should much care where something originates. Just as in the United States we do not ask whether a particular product we buy is manufactured in Pennsylvania or Oregon, we would not care whether the product is produced in the United States, Brazil, Japan, or France. But this is not the kind of world we live in. Even if it were, in a totally free market world the ownership of resources and the decision-making power would gradually move to the nation that had technological superiority.

Thus, if the American people prefer foreign designed and made articles, American funds will flow to the sources of those goods. If American products are regarded by the receivers of our dollars as expensive and inferior, what will they do with the dollars we spend with them? Presumably, they will continue to be willing to accept American money only if there is something here to buy or if they can trade the dollars to still other nations who find our output attractive. Of course, they might buy up our land and natural resources. They might create factories here to produce more goods by their methods and under their managements. Presumably they would realize a high enough return to justify the investment if we were willing to work for lower wages. And we might choose to do just that if we find we lack better alternatives, the inevitable unemployment yielding us even lower living standards.

We are not going to have a totally free market society in the decades ahead. We should assume a world with unsettled disputes, loaded with dangers of economic instabilities, social chaos, undigested change, and even wars. Our ideas about American organization and decision making should be based on the concept that the world will continue to consist of separately controlled geographic and social entities not totally cooperative with each other. In that real world, technical products originating elsewhere bring us both benefits and difficulties. The same is true, of course, about products from this country sold elsewhere. Aside from exports and imports, American corporations own billions of dollars worth of operating assets abroad and many financial and tangible assets in America similarly are owned by foreign entities. Success in obtaining return on investment in trade and in both domestic or foreign-based operations depends for all companies, wherever based, on the policies and actions of the involved national governments. Let us look at how our government's policies influence the competitive position of American companies in world competition, starting with the United States domestic market.

The day when American technological corporations could take for granted great advantages in their home market over foreign competitors has passed. Whether we speak of the entire world market or our own market alone, American technological corporations now are faced with strong foreign competitors. The competitive battle is greatly influenced—and sometimes decided—by the degree of cooperation between a country's government and its private enterprise sector. Unfortunately, as we know, the United States' government–industry adversary relationship militates against victory by American corporations in world competition. Even in competition with foreign companies for the United States domestic market, our complex democracy is not serving us as well as it should. America needs to adjust its pattern of rules and policies for the real world of international competition.

For instance, as each nation goes about trying to handle its domestic problems and relating its activities to the rest of the world, its government is very conscious of the importance of the private sector in technological transfer, either through the moving of products, funds, and national resources or through the management of activities in other nations. Multinational corporations are very influential in technology transfer, specifically in determining the geographic location of investment of private funds in advancing technology and in arranging for the availability of technological products and know-how around the world. Naturally then, each government views it as one of its missions to control the actions of multinational corporations, both those based in its own land and those headquartered in foreign countries. The various governments' approaches to this problem vary, as might be expected. Generally, there is an enormous difference between the United States and all of the other developed nations in the manner in which the federal government relates itself to its multinational corporations.

Other competitive nations appear to have focused on the idea that the way for them to prosper is to win battles for technological superiority. One way they think they can help cause that to happen is to arrange for teamwork between their governments and their private industries. Their goal is that their companies should command the largest possible share of the world market for their products. As a corollary to this, those governments view American corporations quite differently from their local companies, looking upon our corporations as aliens and competitors that are to be handled with adequate bias to ensure the arranging of advantages for the home team. The United States, in contrast, makes less distinction between American and foreign-based enterprises manufacturing or selling products here. Corporations of any origin are all regarded as adversaries. Foreign governments are not much interested in what their companies do away from home. But the American government cares a great deal what American corporations do anywhere in the world. The American corporation is even regulated by the American government in its activities overseas where our government has no direct power to influence the business, social, or physical environment.

Government policies in Japan regarding science and technology are very directly tied to economic growth. They emphasize education in science and technology and arrange to provide the necessary future industrial manpower. The government sponsors technological development most influential in the consumer and export market rather than that which will build national prestige or military strength. In Germany, the policies are chosen to improve the climate for innovation. The government shares costs with local industry in key technology areas or offers interest-free loans that are forgiven if the research and development projects financed by the loan turn out to be failures. Venture capital for new local technological ventures often is provided by a consortium of banks with government guarantees. The Germans deliberately try to establish an environment in which industry will press the outer limits of technology. For instance, industry and government cooperate to set goals for "vehicles of the future." The goal of both German government and industry is to create the greatest possible advantage in the world market for German automotive suppliers. French policy encourages mergers as a route to creating strong national firms where large-scale financing and operation are perceived as being required for success.

Compare these government–industry relations with the situation between our government and Detroit. Regulations as to air pollution and safety are set by government agencies that are not at all concerned with a car's marketability, especially in foreign countries. The government-mandated MPG standards for American manufacturers are based on average performance for their full lines. Thus, a producer that does not sell as many small cars as would be desirable to meet the government's prescribed standards for the average of its cars is not legally able to sell enough intermediate and larger cars to fill the full market demand for them and instead must ration their sales. The resulting cash flow problem is then very penalizing, just when the American manufacturer must re-tool and innovate to bring out new smaller cars to compete with foreign cars that have been smaller all along and have an advantage in volume and cost to produce. In contrast, American manufacturers are operating successfully internationally, making money offshore and losing it at home. This situation suggests that American manufacturers, being international companies, should import small cars and put them through their existing domestic distribution, sales, and service networks, competing against foreign companies whose cars are all made overseas. Then for every small car they sell they could release a larger one, preserving the mandated average. But the government does not allow this—the use of an import to figure into the MPG average of American companies. The effect of the government's mandate is that every small car bought here from a foreign manufacturer displaces an American car.

Other nations seem to have developed a clearer idea of how to relate political policies to technological and commercial trade advantage. In nuclear technology, for example, some years back we sold nuclear plants in many parts of the world with a guarantee that we would supply enriched uranium, which the reactors would need. The enrichment process is a government-controlled area of technology in the United States and, at that time, we essentially had a monopoly in the non-communist world on enriched uranium. During the period when we sold that nuclear equipment, the national policy was that it was a good idea to be engaged in this kind of export. The Carter Administration's policies are to go slow in nuclear energy in general and to be cautious in the reprocessing of spent fuel and the advancing of the fast breeder reactor. This reversal has given us the predictable image of being an unreliable source and trading partner in the nuclear area. No wonder other countries are taking over the nuclear export business. They have developed the capability to reprocess spent fuel and to build conventional and fast breeder reactors. In the last decade we went from having all of the non-communist world's export market in nuclear technology to around a 15 percent share that is rapidly heading for zero.

A specific example contrasts the way other governments work to enhance the efforts overseas of their technological industry, while in the United States the cooperation is often nil and the policies highly uncoordinated. Recently Japan informed Canada that it would not be buying reactors from Canada any longer since, as a matter of policy, Japan had decided to develop its own nuclear reactors. The Canadian government immediately entered the picture to try to reverse the decision. Ottawa used the fact that its significant resource shipments are critical to Japan as a lever to win back a one-billion-dollar order for its nuclear industry. At about the same time, the United States government, fearing nuclear proliferation, stood in the way of American export of nuclear reactor and uranium enrichment technology to Brazil. This handed the Brazilian field to Germany and France. Argentina is pondering whether to buy Canadian-made reactors or German ones; both of those two governments are actively promoting the sales.

Another example of how the United States, unlike competitive nations, fails to support its own national companies in international competition is the $2-billion telecommunications contract granted by the Egyptian government recently to a European group competing against American companies. Although America contributes $1 billion of the $2 billion in total aid Egypt receives each year plus over a billion dollars in military support, the United States government has been unwilling or unable to influence the awarding of contracts in Egypt. Worse than this, the American group found itself facing foreign competitors who were using government financing that was both low-cost and long-term. Our government made no effort to do something similar for the American team. This first contract was only a step in what is planned to be a $20-billion Egyptian program over the next decade.

As to environmental controls, the Administration recently issued an executive order requiring environmental impact statements by American exporters of goods. It is one thing to be concerned about pollution anywhere when a faraway activity can have a noticeable effect on our environment. It is another for us to attempt to police the environmental results of American corporations' activities elsewhere in the world when the same policing is not required of the American companies' competitors in those same countries. We place on American corporations a requirement to meet environmental regulations which are typically more severe than those in the countries where the operations take place. This means American corporations have higher expenses and prices overseas than foreign-owned competitors.

Foreign corporations are permitted by their governments to set up a wide variety of joint ventures in order to prosper domestically and in other countries. Foreign governments very often arrange financing, allow deliberate conspiring to divide up the market, and provide the special advantage of the government's business going to the home-based supplier. They adjust their environmental restrictions with a view toward a balance of developing the market for their home-based companies on the one hand and preservation of the environment on the other. They provide various tax advantages to encourage exports.

In contrast, American corporations are severely restricted in the kind of arrangements they can make with foreign-based companies and foreign governments, even for operations conducted totally overseas. Our antitrust laws follow our corporations wherever they go and these laws create numerous practical problems for our multinational technological corporations. For instance, an American firm may not be allowed to make a foreign acquisition, often the best means of gaining entry into a foreign market, even though the acquired foreign concern is not involved in any way in the United States domestic market. (The rationale is that someday it might be, but the proposed acquisition might preclude such an entry.) A U.S. firm sometimes is not permitted to acquire a foreign concern that is a good source of raw materials or component parts because this kind of vertical integration would be considered an antitrust violation if it took place in the United States. This might apply even though lack of a sure source for these supplies could be extremely penalizing to the firm. In comparison, most of its foreign competitors could buy the same forbidden enterprises with no difficulties with their governments.

What has been said about acquisitions applies equally to licensing and patent protection. American companies may be liable for large treble damage judgments in United States courts for their conduct abroad in restricing others from using their patents and refusing to issue patent licenses to competitors. Such use of patents to provide monopoly positions is legal and open to foreign companies in their own countries. Licensing of trade secrets involves similar considerations. The competitive position of our corporations right here in America is further impaired by an important feature of our antitrust laws, at least as the courts are interpreting them. Almost all American corporate innovators are required to license their proprietary inventions at a reasonable price to anyone who wants them, including foreign companies operating here in the United States. What often happens is that either the Federal Trade Commission (FTC) or the Justice Department's Antitrust Division offers up a consent decree (in lieu of an admission of guilt) that forces the companies involved to license their technologies to competitors, often royalty-free. For example, Xerox was required by the FTC in 1975 to grant royalty-free licenses for three important patents on its copiers, and the FTC set a limit on the royalties that could be paid for the remaining patents. Recently, the FTC demanded that duPont give its competitors royalty-free licenses for its titanium dioxide patents.

Even for the American domestic market, our present antitrust rules are out of date because they do not recognize adequately the enormous countering of monopoly trends resulting from the presence of foreign corporations now doing business in the United States. Many years ago, for instance, it might have been sensible for the government to be concerned about our having a mere three or four automobile companies, with one of them, General Motors, having half or more of the total sales. Now, the large-scale selling in America of Volkswagens, Datsuns, Toyotas, Volvos, Fiats, Hondas, and the rest suggests that no American firm is near a monopolizing of the market. Autos aside, if we should have only one firm in any specific business in the United States, it should not have to be broken up if comparable product lines also are being sold into the United States by foreign sources.

Our antitrust laws, considered against the background of how world trade is conducted, are in some ways naive. For example, the world has seen in recent years the buildup of trading companies and government-sponsored commodity cartels. These worldwide organizations, often condoned or even directed by sovereign foreign governments, are heavily involved in the struggle for control of raw materials and markets for them. They ignore American antitrust regulations, of course, in their global actions. OPEC is an illustration. Another is the so-called "Uranium Cartel" which received a substantial amount of publicity in this country in the late 1970s. To understate the situation, our antitrust rules are being violated every day by OPEC activities which reach into the United States. Similarly, if other nations choose to control the sources of uranium to the maximum extent they can, this affects uranium supply in the United States and the price of uranium for our electric power utilities. Our government's antitrust units can go after our own private companies if they seek to fix prices or to allocate markets in either petroleum or uranium. But we cannot put other nations in jail for violating our laws, especially those countries that have passed specific laws of their own prohibiting such challenges by us against the cartels they favor.

For our companies to be truly competitive with other countries' companies, a reworking of our antitrust legislation is needed. The updated law might recognize that greater competitive strength in the world sometimes results from companies being put together. Selective combining may sometimes create more, rather than less, competition in our domestic market. The government might create an agency — say, in the Commerce Department, in a sort of juxtaposition to the existing Antitrust Division in the Justice Department — to promote and foster sensible mergers. Of course, a merger-stimulating activity lowering competition in the domestic market would be a violation of both the spirit and the letter of existing antitrust legislation and the new agency is not being proposed for such purposes. It would seek always a healthy, profitable group of competitors, in contrast with the frequent situation where the nation has one satisfactory large operation plus a string of lesser ones, all of the latter in danger of going out of business and unable to afford adequate technological development. This kind of action would be especially sensible to consider if all the American-based corporations are unable to compete against strong foreign-based corporations whose governments are solidly behind them. Probably the nation's antitrust laws also should be modified to permit the formation of American global trading syndicates to market our products overseas with no bans on the arranging by competitive corporations of joint ownership in such enterprises.

Foreign-owned corporations operating in this country must pay United States income taxes on their earnings here. However, foreign governments are liberal in allowing credit for these tax payments as they assess taxes on the overall income of their corporations. The matter of tax credits on income earned abroad by American corporations has been a constant problem here—a manifestation of the adversary relationship. Some in America say our corporations should not be encouraged to operate outside of this country and that to do so exports jobs. Accordingly, they think the tax structure should be such as to discourage American corporations from becoming multinational. If income is earned abroad and taxed there, taxing it again in the United States is double taxation. But these critics of American activities abroad believe that is exactly what is needed to demotivate off-shore investment. If foreign governments do not engage in this taxing practice and we do, it helps foreign companies take over the world market.

Soon after World War II, the American market began to be flooded with products from overseas that threatened the sales, earnings, and employment stability of American corporations. Lower labor costs in other countries led to lower-priced foreign products which were favored by American consumers over the higher-priced domestic goods. In figuring how to counter this trend, American corporations recognized they had more assets here than mere production facilities, including broad experience in the United States, a deep understanding of the American customer and market, and a well-honed organization in place for distribution, servicing, and maintenance. They knew their only competitive weakness at that time was the relatively higher cost of domestic labor. It was natural for them to decide to go overseas for some of their manufactured parts and thus be able to hold onto some of the American market they were losing to foreign companies and keep their domestic organizations healthy and growing. In most instances their manufacturing activities expanded here at home, since only portions of their products were produced overseas; critical parts manufacture was kept here along with assembly, test, and domestic distribution. The American corporations merely had the easy choice of either losing the total domestic business or holding onto most or some of it by going overseas for part of their production.

American ownership of overseas production facilities has been and still is opposed by some labor unions. Of course, to be really effective in stopping foreign production from coming into the United States, the act of preventing American corporations from engaging in some foreign production would hardly scratch the surface. Even if American corporations were required by law to keep their technological know-how and products manufacture at home, it also would be necessary to keep the lower-priced foreign manufactured products out.

But a lower labor rate as a basis for overseas production is no longer the rationale it used to be for technological products. European and Japanese labor rates have come up to equal our own or sometimes surpass them. There remains the lower-priced labor of the underdeveloped countries. European and Japanese ownership of operations in those countries presents us with the same problem. Either American corporations have to follow suit and place some production in the underdeveloped countries, or we must enact high tariff barriers on imports into America, an action that would raise prices to American consumers.

Lower-cost labor has not for years been the important reason for our companies to manufacture in other countries. The new phenomenon that has demanded American corporations' interest in operating overseas with technological products is the growing technological status of Japan and western Europe. If American corporations that have a technological lead fail to set up manufacturing operations in other countries, their lead cannot be long protected. Even if initially we are quite far ahead in a product area, and hence it is advantageous for other countries to purchase exported American products and gain economic benefits from their use, similar technology typically can now be developed abroad in a very short time. The world market then is opened to foreign-based corporations. The way for American corporations to exploit technological leads is to establish head-start positions hurriedly in the world markets.

Parts of the United States government have not been oblivious to the fact that a world market, rather than merely a domestic one, must be considered as the competitive battle arena. However, most of our government organizations, laws, and policies do not adequately reflect this. World trade has only recently begun to be perceived by the government as important to American business. Only in 1979 was a proposal made to consolidate all government agencies involved in trade in a Bureau of Trade (within the Department of Commerce). This would give the nation its first ministry of trade, an agency of the type that other industrial nations have had for many decades.

To win world markets we must send Americans to work overseas. This requires that they be given special financial inducements by their companies to compensate for taking families abroad and to cover unusual expenses (relocation, higher living costs, and the like). However, our government has been anxious to so interpret tax rules that everything a corporation does for its overseas employees is subject to full income taxation. The United States is the only major country that taxes its own citizens for their special overseas location expense reimbursement as if it were ordinary income. By taxing the reimbursements for extra expenses, the government makes our corporations' overseas representatives' taxes higher and our companies' costs higher as they try to compensate. This discourages American companies from seeking the world market. Because of these tax policies on personal income earned abroad, our business entities are greatly handicapped in important competitive business opportunities in the world market. For instance, the United States is now fifth in overseas construction. A mere few years ago it was first. Large construction projects lead to large heavy-equipment exports. They added up to $75 billion for 1978, but are declining now because of severe competition. For this area of export business, it is vital that our engineers be located in the foreign market regions, drawing up the plans and specifications and calling in those documents for American equipment.

Investment to advance technology domestically is greatly influenced by our tax policies, as we have discussed earlier. This applies also internationally. Investment by American corporations overseas seeking to exploit technological leads can be encouraged or discouraged by our government's tax rules on overseas corporate earnings. The records show that the cash flowing back to the United States in dividends, interest, and royalties has been greater year in and year out than the dollars sent abroad for investment purposes. These dollars returning to the U.S. help to solve our difficult problem of a balance-of-payments deficit. Moreover, they constitute a supply of capital for investment domestically in further technological advance and a source for the stimulation of the domestic economy such advance would bring about. A survey of American corporations, separating them into those that are active only domestically and those that have international operations as well, discloses the latter group to have expanded employment in America more. By playing the world market, the multinational American corporations have done better at creating job opportunities here at home than their counterparts that avoid foreign operations and are exclusively domestic companies.

The combination of returns on investments, revenues from exports, and payments received for licenses and technology transfer have added up on the whole to the betterment of the returns for shareholders in American corporations. Through an increase in domestic employment, they also have worked to the advantage of American labor. This is in part because a successful American corporation that operates abroad as well as domestically usually increases its exports based on production at home. It sells overseas more specialized components that are made here and then assembled in foreign countries, together with other foreign-manufactured parts, to complete the product to be sold there. This apprach has given our companies larger shares of the total world market than they otherwise would have had. An American corporation that limits itself to the domestic market tends to be smaller in its overall activities and less perceptive of technological and market changes. If it does not lose out to another American corporation that is broader through its international operations, it gives way to a foreign-based company active here and throughout the world.

Despite these arguments there is concern in the United States that when American companies make overseas investments, jobs are exported and unemployment is created here. This result, combined with the always present tendency on the part of the government to look for additional revenue sources, produces the constant proposal that earnings on foreign investments should be taxed more severely. Of course, to do so would greatly curb such investment. It would also mean reduced capital availability in the United States stemming from the return on investment and would handicap American-based technological corporations in their competition with foreign-based ones. As it is now, the government does tax earnings of United States foreign subsidiaries when they are distributed back to the parent American companies. In Switzerland, France, and other nations, such taxing is non-existent or else is very small. Most other countries seem to recognize that operating internationally is helpful to their domestic economy. They encourage it and they avoid taxing policies that would discourage the return into their countries of funds earned abroad on good investments. The United States practice is to avoid double taxation by allowing a credit for taxes paid on those same earnings in foreign countries. However, the policy of taxing fully earnings that are repatriated naturally causes American-based multinational corporations to avoid that repatriation. It is better, for example, to borrow domestically if more capital is needed here in the United States than to bring back funds from foreign earnings, sharing these funds with the federal government.

In effect, the government blows hot and cold on foreign investment by American companies. The net effect of present policies is to handicap the optimizing of technology leads. This holds back the creation of further investment capital, which the fullest exploitation of U.S. technology would provide. Hence, new technology leads, new product developments, and the additional jobs that would go with such investment are lost.

Considerable influential opinion exists in the United States that we are allowing our technology to leave the country, to our disadvantage. More particularly there is in many quarters the feeling that we have built up and are still aiding foreign competitors who later take markets and jobs away from Americans. Still worse, it is believed by some that our generous and loose technology transfer policies and habits are of material assistance to potential military enemies. What we have done and will do about the technology underlying national security, and our policies on international technology transfer, are important factors in our technological status. These issues now deserve further discussion.

15

national security and technology transfer

In an earlier chapter we described how important military technology is to the overall status of American technology. Of course, the strength of our military technology goes very directly to the substance and credibility of our national security. Also, the size of our military technology budget has a great deal to do with the way all technological resources end up being allocated both by government action and our free enterprise system. Thus the military technology programs affect our non-military technological efforts beyond the matter of similarities and spin-offs between the two sectors. Our international economic competitiveness in matters technological is influenced by the nature and size of our military technological projects. One way or another, what we do on the military technology front affects our political and social stability, and these issues turn right around and exert a control on the rate of our technological advance and the status of our technology vis-a-vis other nations.

That a large part of our total R&D effort is devoted to military weapons systems is an important parameter in government–private sector relationships. If we take into account all of the factors that go to forming our military technology activities and then bring in all of the interconnections with the non-military side of our nation's endeavors, a grand sphere of policy is seen to be involved. The government, not the private sector, has to be at the center of this orb or its segments will not hold together.

In this chapter we are going to explore further this important interaction sphere, considering first some military technology policy questions. We shall also find it advantageous to connect our discussion to another broad issue: technology transfer from the United States to other nations. International technology transfer out of this country is itself a large and perplexing issue for our policy makers. It arises quite obviously in connection with necessary attempts to control technology shifts to potential enemies, but the issue is important as well if we look at it only as it bears on maintaining significant American technological superiority as an essential base for foreign trade balance and economic security.

The military technology programs of the United States over the past several decades have advanced technology enormously on several fronts. We described earlier how aeronautics, space, computers, communications, materials and processes, nucleonics, and numerous aspects of the physical and biological sciences have been pushed across new frontiers by the many billions of dollars of R&D funded by the Defense Department. This broad technological expansion has had an accelerating effect on the advance of non-military or commercial technology. However, the fallout has not been so great as to suggest that for every dollar of military technology expenditure we realized almost as much advance of the non-military fields as if we had spent it directly on civilian technology. Probably our relative productivity increases and our net rating in technology vis-a-vis other nations have on the whole been hurt rather than helped by our heavier involvement in military technology as compared with other nations.

In the past thirty years, had the total dollars we spent on military R&D been expended instead in those areas of science and technology promising the most economic progress, we probably would be today where we are going to find ourselves arriving technologically in the year 2000. Not everything in science and technology could have been accelerated by stronger assignment of resources because some breakthroughs and their follow-ups simply must happen in series. But for most developments, working a problem directly gets us farther along than merely waiting to get to it or helping it with fallouts from other projects. We have not had the kind of world in which we could ignore military technology requirements. Technology programs for defense have been and continue to be necessary. Military technology superiority, it appears, will continue to have a great deal to do with our ability to so influence the world that our own way of life, freedom, and security are maintained.

Maybe we have had about the right emphasis on military technology development and maybe not. It is impossible to know for sure. However, we can say that a distorted allocation of technological resources toward military weapons systems can hurt the overall economic and technological stature of a country. If our military expenditures are greater than they should be relative to other nations allied with us, our standard of living is hurt by comparison with theirs because those expenditures subtract from, rather than add to, what is produced per capita for civilian use. Capital investments made on behalf of military R&D and production leave less capital to invest in non-military developments. The employment of a large fraction of the best scientists and engineers on military projects means they are not available to advance the store of knowledge and innovate along non-military lines. Our disproportionate share of the military weapons requirements of the non-communist world has accordingly handicapped us by comparison with our industrialized allies.

Germany and Japan have been extremely successful in their competitive technological endeavors in non-military fields. Over the past ten years, the United States has expended two or three times as large a fraction of its GNP on defense as have these nations. In fact, Japan's military activities have cost less than one percent of its GNP. Neither Japan nor Germany has a nuclear capability in military weapons systems and not a yen or a Deutschemark has been spent to develop one. Japan has a tiny military force and Germany's contribution to western European defense on a per-capita basis is less than ours, yet the overall per-capita incomes of both nations have more than caught up with ours, everything properly considered. Those two nations have a low inflation rate and have stood for many years under an American protectorate umbrella in military matters.

If there is a danger to Japan, or a danger to us through Japan from some third power, it presumably could only be the Soviet Union or China. A small nuclear bomb stockpile created by Japan, with a correspondingly small, ready force of Japanese intercontinental missiles, should be sufficient to deter both of those potential enemies from attempting a military invasion of Japan. Most of those who make it their life's work to study such problems are quick to point out that both the Soviet Union and China have other problems and opportunities better deserving of their attention than an attempted military takeover of Japan with the risk of nuclear weapons coming back at them from Japan. In any case, the idea that we have an obligation to provide an umbrella of protection over Japan, enabling her own military endeavors to be at a minimum, may be out of date.

When America takes on itself too great a share of the defense burden of the non-communist nations, the structure of our economy can become overly dependent on military budgets. For instance, it can be noted and should be a matter of concern that in the United

States whenever a major weapons system begins to be seen as of doubtful necessity, its proponents immediately argue it must be continued or otherwise unacceptable unemployment will result from cancellation. It is true that a short-term dislocative increment is thrown at the economy with a sudden halting of any major military project. Surely, however, continuing a program we don't need is "no way to run the railroad." Carrying on with military programs that are no longer suitable, purely as a make-work activity for the economy, is the equivalent of running tracks out to the wilderness to keep the railroad workers busy.

What we would like is a sensible balancing of security risks against costly allocation of our resources to military preparedness. We would like to realize this proper tradeoff not only for the United States, but for the United States in concert with its allies. Unfortunately, the weighing of security risk against the size of the military budget is at least as tough to accomplish as the balancing of environmental protection and safety relative to benefits from technological developments in energy, food, transportation, and other areas of technological advance.

In fact, in these other areas, in contrast with the military, the balancing problem is easier because we are better able to list the benefits and the costs to obtain them. True, some risks cannot be judged because we lack the knowledge, as in assessing the effects of minute amounts of radiation. However, in military matters we have to guess, first of all, about potential enemy action and objectives. Then we have to work out the complicated sharing of the responsibilities and costs with our allies. This is not easy because their evaluations of the tradeoff of security risk taking versus costs are different from ours. Next we have to recognize that security is not based on military power alone, but also includes national morale and solidarity, economic strength and competitiveness, diplomatic skill and strategy, energy resource availability, and numerous other influences along political, social, and psychological dimensions. Determining the size and nature of our military programs in association with allies thus involves many more intangibles, unknowns, and complexities even than deciding on nuclear energy installations.

In military technology, we do have one powerful advantage, however. The organizational problem of roles and missions is easier. Our accepted system is already probably the best system for us. The President, as Commander-in-Chief, has clear responsibilities for proposing military policies and programs. The Congress has an established mission to approve and provide the budget for the programs proposed by the President and to debate and help form the complex interface of military versus other dimensions of security and our relations with other nations. Full-scale inquiries and deliberations occupy the attention of the Congress at least two or three times a year for budget approval (and additionally for such special issues as SALT, the Panama Canal, Near East crises, or actual warfare breaking out). To be sure, all of the discussions are heavily political, which means they are only partly objective. Numerous shortcomings and doubts about our approaches are continually turned up. It is an imperfect system. But it is not evident that substantial organizational variation would help in this area.

Our military technology programs are not suffering greatly from a critical problem we have called out repeatedly in this book, with regard to civilian technological advance: the government–free enterprise relationship. In military technology advance, we have found a way for the government and industry to cooperate. Military requirements are set by the government; industry is involved sensibly, to provide specialized studies showing what technology can do. The government chooses weapons systems to satisfy the requirements it decides upon. To get the weapons designed and built, it contracts through competitions with the industry, where the capability resides. This procedure is not flawless, but it works. If our engineers were smarter or our military able to anticipate more brilliantly the relationship of military requirements to technology, and if we could improve the efficiency of the Defense Department and of the military contractors, we could get more for each dollar expended. But we have been striving in these directions for some decades. While somewhat better performance can be hoped for, vast improvements are not foreseeable.

Military R&D is not without its handicaps. They stem from interservice rivalries, military–civilian viewpoint differences within the government, and more than a mere nuisance effect of political constituencies' constantly pressing for special benefits through the winning of military contracts. Military technology advance also is slowed by the familiar problem that a spread of value judgments exists on the part of all influential groups as to risk versus costs versus security benefits obtainable. Because it is inherently international in nature, and depends on strategies that in turn rest on unknown actions of potential enemies, a military technology program must be based partially upon arranging a consensus on strategy and implementations with our allied nations. The real bottleneck to technological advance lies in these non-technological aspects. It is important to our security that the plans we lay out with our allies employ technology, economic strength, and political skills in a harmonious, optimum way.

When it comes to military technology, our democracy does have a system for decision making and implementations and we have solved the problem of government–private roles. However, we are terribly weak in handling the broad systems problem: the balancing of risk to our security versus the economic and social costs of efforts to reduce that risk, the tradeoff of assignment of technological resources versus the added security we gain. We need more emphasis on this balancing problem because in America we tend to approach security policy by short-term, simplistic politics, for instance, by dividing into two extreme camps, the hawks and the doves.

From the hawks we always hear that our military program should be larger. The tendency is to compare our expenditures with those of the Soviet Union and our array of weapons against their inventory. If we are behind or have smaller efforts in any known respect, the recommendation from the hawks is to accelerate and enlarge our programs. A B-1, an MX, a new aircraft carrier, a neutron bomb, a cruise missile, an anti-ICBM, a killer satellite concept, or a reinstituting of the draft—the hawks are always for it. On the other side, we have the doves, who constantly remind us that weapons can only kill and money spent on weapons should be spent to solve social problems instead. The doves always press for a smaller defense budget.

Neither of these groups is heard very often discussing the more difficult, sophisticated, real problems of balancing risk versus expenditures and military versus non-military approaches to security. Although each group, if pressed, would make token admissions that these tradeoffs are fundamentally where policy should start, they are little interested in lingering at the starting point to choose a direction before running off. Thus, that our technological resources are limited, that investment in military means less investment in non-military, that the world is dangerous, that potential enemies do exist, that security depends also on economic strength, social stability, and diplomatic skill, that inflation is a serious problem in setting an economic base on which military security and all other dimensions of security can rest, that reducing government spending is an important parameter in the inflation-halting process—interconnections such as these, linking military considerations with other issues, receive inadequate attention.

Let us illustrate this with the example of NATO. A group of nations on each side of the Atlantic is cooperating to deter, or be able to counter if the deterrence fails, a possible Soviet Union attempt to take over western Europe by military force. If there is to be an invasion it presumably will be because the Soviet Union has decided on it. The Soviets will have the advantage of the offense and of integrated control by a single nation. Opposing them will be a cluster of military forces from over a dozen autonomous nations who merely cooperate. A central NATO executive control exists, but it is rather loose since the various nations are independent and provide as they see fit for their armies, navies, and air forces. NATO headquarters has no power of decision over the separate national forces and their procurements. The Soviet Union has more soldiers, planes, tanks, and guns in Europe than the combined western forces.

In certain areas of technology, the NATO nations are superior to the U.S.S.R. This is particularly true in advanced electronic techniques that go to command and control communication, intelligence, and reconnaissance. Unfortunately, although their information technology is ahead of the Soviet Union's, the NATO countries have coordinated and unified their efforts in this field only modestly. They have not used fully their trump card of superior information technology to make possible so focused a military response to an invasion as to give NATO the equivalent of a stronger force even though it is the Soviet Union that has the advantage of the offense and quantitative superiority in men and equipment.

But there is something else that makes reasonable the questioning of the present credibility of NATO. Much of NATO's military potential rests on the idea that the United States will use nuclear forces if necessary to halt a Soviet Union military takeover of western Europe. Many years ago the threat of use of nuclear weapons by the United States in western Europe to counter a Soviet invasion might have had high credibility. Today, is it reasonable to expect the United States would use nuclear weapons to stop a Soviet Union military onslaught across West Germany and on through the rest of western Europe? If we were to drop nuclear weapons on Soviet soil, we would be inviting them to send nuclear weapons to America in return. Can we really expect the western Europeans, particularly the Germans, to be happy with the idea of our resisting an invasion by U.S.S.R. troops by setting off nuclear bombs in their path? With even the most optimistic impression on the part of western Europeans about the fineness of accuracy and control of those weapons to minimize radioactive fallout and concentrate the damage, would it not look like a way to inflict enormous, unacceptable damage on the countries the Soviet troops move into?

It is to be expected that before the Soviets would launch a western Europe invasion they would force the matter of use of nuclear weapons there to be discussed in America. They need only make noises about imminent war in western Europe and tell us many times over that for us to try to use nuclear weapons would force their reciprocal use of them on us. After such specific threats, how many days of discussion would it take before a consensus would form in America to preclude the American President's allowing our nuclear weapons to be used in Europe unless the U.S.S.R. uses nuclear weapons first?

The body of nations making up Western Europe has a total population, industrial strength, general wealth, and technological capability not too different from that of either the United States or the Soviet Union. Even without us, if they have a serious interest in stopping a Soviet military invasion, they should be able and willing to create a credible defense. If America were to take itself totally out of western Europe — troops, nuclear weapons, and all — the Germans would have to and could create quickly a nuclear weapons system capability of their own. A German leader then could inform the Soviet Union that an invasion of Germany would bring nuclear weapons onto Soviet cities. The Soviets, of course, could counter by saying they would utterly destroy West Germany if nuclear weapons were employed by the Germans on the Soviet Union. But in the trade the U.S.S.R. might lose Moscow, Leningrad, Kiev, Stalingrad, and a substantial part of all Soviet industry. Contemplating this possibility might be enough to cause them to rule out invading western Europe and to propose instead a substantial degree of mutual disarmament that would spare both sides the economic drain.

Many experienced observers of the situation in Western Europe argue that the Soviet Union has better ways to arrange for its security and the meeting of its world ambitions than to launch a military invasion of western Europe, with the immediate task of policing it should they conquer it. With less expenditures of all kinds and a greater chance of success, they can encourage a gradual changeover of the society of western Europe to one that more closely resembles theirs. This might realize for them about as much cooperation from western Europe as they could expect to get if they became military occupiers over a few hundred million people who would loathe and be determined to oppose them.

Even if the United States does essentially nothing to change its policies, Germany and Japan and other nations soon may see good reason to want to change theirs. They have to be fearful of relying so heavily on America and on American nuclear weapons in particular. President DeGaulle came to this conclusion many years ago. He decided it was unwise for France to depend upon American nuclear bombs to keep an enemy out of his country. Instead he arranged for his nation a small (but highly impressionable, so far as the Soviet Union is concerned) nuclear weapons capability of its own.

The Soviet Union has the ability to knock out western European cities with nuclear weapons and it is increasingly in a position to threaten western Europe militarily to force compliance with its policies. It need have little fear of retaliation if, as a means of saving western Europe from Soviet attack, America connot be taken seriously as ready to send nuclear weapons against the Soviet Union and receive similar punishment in the United States. Western Europe cannot be blamed for exhibiting apprehension about American nuclear employment as a protection barrier. This is indicated by the lack of enthusiasm with which NATO nations have been responding to the recent proposal that their countries be bases for our short-range, nuclear-tipped rockets, intended as a counter-threat to the Soviet Union's similar missiles. West Germany appears willing to house some of these rockets provided other NATO countries will do the same. The Scandinavian allies refuse to accept nuclear weapons, while in Italy, Belgium, and Holland the governments have such slim political majorities sympathetic to housing such rockets that opposition seems as likely as acceptance.

With the foregoing points about military technology as a background, let us now take up the question of technology transfer from the United States to other countries. The first reason why restricting the transfer of technology must interest the United States is national security. That must take precedence over the matter of technology transfer's impact on general economic competitiveness. The fullest use of scientific and technological advance on behalf of the world's society, if we rule out security matters, would appear to require that the know-how of which these tools are comprised be freely transferable about the world. Just as world trade is enhanced by free movement of money, manufactured products, and raw materials, so the mobility of science and technology affects the benefits obtainable from these tools. To a technological corporation the greatest return on available technological resources comes from that corporation's having the total privilege of moving its technological ideas, overall know-how, and actual products from its various operations in the countries where it is active to foreign companies and customers and even to governments in those countries wheie the governments have control of the technological applications.

However, governments restrict technology transfer at both ends of the transfer process. In the United States, apprehension and indecision characterize the government's effort to determine what should be the proper rules and controls for technology transfer to other countries, whether actual products or technical data. The government is involved because technology is judged to be an important asset whose shift to other countries can affect the general health and security of the nation. Of course, some of the advanced technology is directly a result of classified R&D supported by the government to provide a base for American security and, more specifically, for the design of American military weapons systems. The government must be the arbiter as to what is militarily classified and who, even within the body of United States citizenry, has a need to know and should be permitted access to the technology. It goes without saying the government should ban the transfer of classified technology to other nations except for deliberately planned military cooperation, and then the transfer has to be under strict governmental control.

The problem is more complex for technology bearing on national security but not classified. The specific technology may or may not have arisen from government-sponsored military effort. The question is whether the security of the United States would be impaired by allowing the technology to leave the country no matter how the technology arose. This question in turn suggests others. For example, does not any technology shift to other countries have the chance of aiding potential enemies? If so, should we not stop all exports of advanced technological products and try to put a barrier on any leakage of new science and technology information from the United States?

An extreme, total limitation would obviously be ridiculously impractical to police. What constitutes a transfer of technology is in many instances impossible to define. If we were to seek to prevent export of any technical knowledge for fear it will somehow be used against us by an enemy, we would have to take ourselves out of technological world trade. Semiconductors and computers would be banned as export candidates and we would not sell commercial jet airplanes to foreign-based airline companies. Severe restrictions, even if less than a total ban, would put a handicap on American technological corporations in overseas trade. Almost everything that takes place in the United States on the technology front will find its way to other countries in time, if indeed we start out ahead. We would have to create a crazy pattern of living to try to restrict the ability of foreigners who come to the United States to observe what we are doing. We could hardly prevent their purchasing and studying our non-military, commercially available technological products.

We would have to deny to our own scientists and engineers the privilege of educating each other at their public professional society meetings because they might be overheard. We would have to stop our technical publications about new science and technology because the reports might be read by foreigners. This would put an absurd, fatal constraint on the functioning of the entire scientific and engineering establishment in the nation and probably it could not be caused to happen in toto anyway because of the First Amendment and other constitutional constraints.

Still, if certain critical technology items are highly advanced, why should we allow a potential enemy to have them or knowledge of them, saving themselves even a year or two? Should not some advanced aspects be withheld at least partially, wherever we actually can do so? For instance, why should the Soviet Union be allowed to buy one of the fastest and largest computers ever built in the United States if the only application we have ever been able to find for that computer, and the reason it was built, was to aid in the design of nuclear bombs? As another example, if we develop techniques for forging materials of such unusual composition that they make possible higher temperatures, efficiencies, and thrust-to-weight ratios in our jet engines, why let out the details of this process to the U.S.S.R.? We can hardly be sanguine about supplying high-technology products to the Soviet Union on the assumption that these are "non-military" and would have nothing to do with their defense strength.

Some questions about how far to go in embargoing transfers of technological equipment to the Soviet Union can be answered quite readily with simple common sense. For instance, some things we produce that the Russians might choose to buy from us are also offered by other nations with approximately equal quality, delivery time, and pricing. Given a choice, the Soviet Union's hard bargainers might select an American article, preferring it over products of other nations for only the slightest differences in performance–cost–delivery. Obviously there is little point in refusing to sell such articles to the Soviet Union. A ban by us would not deny them the article but would merely shift their purchases to our competitors and hurt the American supplier.

Moreover, if the difference between an American product and one available from another country is more than just slight, it would be natural for the Russians to seek to obtain the American product through others. To stop this, we would have to arrange, every time we sell one of those pieces of equipment outside of the country, that it does not get transferred to the Soviet Union with only trivial delay and some slight added cost to them for a commission in the transfer.

Usually, it would be totally impractical for us to attempt to control all resales. There would be no sense in forbidding sales to the Soviet Union unless we have the full cooperation of other nations to whom we are willing to sell and a firm understanding they will ban resales. Without such commitments, we would have to end up banning sales of American equipment to these other countries as well. For marginal penalties to the Soviet Union, we would rarely want our export business to take the beating such broad embargoes would represent.

Another question, which since the Soviet invasion of Afghanistan looms more important than before, is how to enforce the rule that equipment sold to the U.S.S.R. for one purpose must not be used for another. An example is a widely described case of a computer made in the United States and sold to a West German subsidiary of a Swedish company for resale to the Soviet Union. The understanding was they would use it in design of synthetic rubber. Not far from the design agency for the synthetic rubber industry of the Soviet Union is the Tupolev aircraft plant that makes military aircraft. The American manufacturer of the computer pointed out that the machine, when sold, incorporated technology already eight years old, with a processing speed fully in conformance with the United States restrictions on computer exports. The sale conditions required that the American company's technicians be permitted periodic inspection visits to the machine in the Soviet Union to assure its continued use for the purpose for which it was sold. Furthermore, some American experts have said that the kind of calculations needed for aircraft design are essentially beyond this machine's capabilities. However, most technological sophisticates in the United States would argue that it is unimaginable we could police the utilization of computers, or any other kind of equipment sold for one stated purpose, to be sure it is not utilized for something more directly of military advantage to the Soviet Union.

Of course, the silliest kinds of things to deny to the Soviet Union are those items of equipment sold commercially in the United States to anyone who walks in the door with the selling price in hand. It would be absurd to set up a system to deny direct purchases by visiting representatives of the Soviet Union, because it would be a cinch for them to arrange that someone else in the United States purchase it and resell it to them with no more than a minor commission (or a bribe, if we pass laws against resale). What could work for a multimillion-dollar computer or a jet engine or an order for a thousand submergible pumps for oil wells obviously would not work for widely sold, mass-produced laboratory instruments or large-volume small computers or even diesel truck engines.

American and European firms provided much of the technology for the U.S.S.R.'s truck production at Kama River. This plant produces a quarter of a million large trucks and a third of a million diesel engines annually, as well as other vehicles. The output of this plant, thanks to exported western technology, is superior to the standard truck now used by the Soviet army. The foundry associated with the facility, the largest of its type in the world, is run by an IBM computer. The plant produces tank parts, even though assurances were given originally that it would produce only civilian vehicles.

Similarly, the United States sold the Soviet Union precision grinding machines for the producing of miniature ball bearings, one application of which is in missile guidance systems. Today the United States dominates world production of oil drill bits, producing several hundred thousand units per year. Designed with American aid, a Soviet plant will soon have the capacity to produce a hundred thousand units, a figure exceeding the Soviet Union's estimated internal requirements through the decade of the 1980s. It can be certain they will enter the world market for oil drill bits and be a competitor. Aside from this, if the Soviets were not able to increase their internal energy supply economically, they might have to go outside for their purchases and the consequent financial strain might limit their spending on military weapons systems.

Usually, denials by the United States government of specific transfers to foreign countries, wise and justified as they may be, can have only a temporary effect in our maintaining technological leads. As new techniques find a place in commercial product applications, it eventually becomes impractical to control and halt the export of such technology. Even with security strongly in our minds, we must take some risks and allow the work of our own technological world to proceed. The Defense Department has recommended that the know-how embodied in the most advanced and critical technology be highly controlled as to transfer out of the country while export of finished products might be handled with no controls, or less structured ones. But no clear and binding United States policy has emerged. In fact, when Soviet trucks and tanks moved into Afghanistan and America, from the President on down, developed a new interest in the Soviet Union as a potential military foe, our policies regarding transfer of American technology to the Soviet Union simultaneously also became subject to broader scrutiny. Most people had assumed it always had been our policy not to sell to the U.S.S.R. products that could help the Soviets militarily. Yet, suddenly, President Carter ordered an embargo on such transfers, which tended to suggest to the public that we had been engaged in them until the Afghanistan invasion.

Numerous disclosures then appeared in the media, one about the advanced equipment that had gone to the Kama River Plant producing trucks and tanks, another about financing by the export–import bank at the level of hundreds of millions of dollars, an important factor in enabling the Soviets to purchase these American-made items. Accusations were voiced that the procedures used by the government to approve individual items of technology export were being kept secret, that the Russians in the past had been given favors in return for political or semipolitical cooperation elsewhere in the world. The House Armed Services Subcommittee charged that the Commerce Department's Office of Export Administration was not even enforcing its own regulations, which prohibit the military diversion of U.S.-licensed technology as appeared to be happening at the Kama River Plant. In panic an official in the Commerce Department suggested a license be required before any American could discuss advanced technology subjects with a foreigner.

Whenever the technology is classified, the government is more readily equipped to choose what areas to control and how to organize to achieve that control. When it is not, the decisions on transfer could be very lenient, with exceptions specifically called out by the government. At the moment, for example, the areas under restrictions might include new developments in semiconductor technology and the most advanced aspects of computers, such as microprocessors (tiny computers on a single chip). At any one time the number of such exceptional areas might number no more than a dozen, so the administrative chores and delays in joint consideration by government and private industry of the transfer process need not be unreasonable.

Unfortunately, private industry cannot be relied upon always for transfer actions in the total national interest. For instance, a new small company based upon advanced technology and having difficulty financing its first several years might find it very attractive to accept financing from abroad on liberal terms. That company might be in no position to consider deeply the long-term benefits to the nation, even if it presumed to know what those benefits are, because it must solve an immediate financing problem. Rather than losing everything in the short term, the company may elect to gamble as to the long pull and allow its technological lead to be shared with foreign companies.

The best way to control technology transfer from the United States so that the working out of all factors will be in the national interest is to have, in the first place, a strong economy and an environment favorable for investment in new technology. Solving our basic problem of inflation, for example, and increasing the cash flow of corporations to make possible more speculative technological investment can keep us ahead, and staying ahead is the best way to protect ourselves against foreign competition. Sending technology abroad may prove somewhat harmful later on, but if we are paid well for it in the short term, and if we can remain substantially superior to competitors, no bad long-term situation will develop. We can exploit short-term situations one after another until the whole becomes a long-term success. Even in competition among American technological corporations for the domestic market, staying ahead of others by keeping one's know-how secret is generally not as likely to yield long-term success as continually generating useful secrets faster than the competition. True technological superiority is far better than manipulation of the administrative controls on technology transfer.

All in all, the issue of transfer of technology from America, including its transfer by American corporations setting up operations overseas, is a political issue that is undecided. It will probably remain in dispute for years because it is a "good" political issue. With substantial anti-business sentiment in the United States, some politicians can make points with their constituencies by arguing that what American technological corporations do in technology transfer must be carefully watched and regulated. If not, it is argued, the American citizens' interests will be impaired as the corporations pursue paths that selfishly optimize only their shareholders' benefits and not those of the nation as a whole.

The free enterprise role in technology transfer would be simple if we had a true free world market. We then would simply allow the various companies that are seeking to derive a good return on investment to go about their separate optimizations. The individual transactions involving technology transfer, long-term or short-term, would be vast in number. Some would be made with lack of adequate competence and knowledge. On the whole, however, as with free market operations in general, the private optimizations would work out better than any attempt at government control. However, we do not have that kind of free market. We have security issues that must be addressed, and even for non-military economic interchanges among the nations, government interference exists on almost every front. The governments of all countries of the world seem to have concluded that the free market acting alone will not provide them with the best match of technological development to their countries' goals. In addition to their being engaged in some form of sponsoring of technological advance, those governments are usually involved in regulation of technological applications and in controls on technology transfer.

It may well be true that the United States erred in allowing too liberal a transfer of technology during the period from World War II's end to a decade ago, at a time when our technological superiority was manifest. But realizing that does not tell us exactly what to do now. To complicate matters, restrictions on technology transfer work in both directions, and we must consider the attitudes and controls of the governments of other nations and not those of the United States alone. While we may lead in the science and technology olympics on total points, we cannot win every contest today. Especially as time goes on it will often be other nations of the world that will make the discoveries, do the inventing, or create the proper follow-through to application and market development ahead of the United States. We have already said that one benefit to an American technological corporation from operating a unit elsewhere in the world is that it can take advantage of the technological developments and market trends that exist outside of the United States.

An American multinational corporation becomes a participant in the scientific and technological life of the various areas where it operates. Its foreign employees in those areas will make important contributions. It is in the national interest for the parent corporation to be able to move the technology from wherever it arises back to the United States. However, we should not be surprised if the governments of the various countries where the technology originates want a say in the transfer of that technology, for the same reasons we listed earlier for the United States. For economic competitiveness or security reasons or protection of jobs at home, those other countries will present obstacles.

Restrictions on the flow of technology from the United States to other countries, even if at least partially in the national interest, do create for us a competitive disadvantage in exports in the short term. To try for careful and thorough investigation before granting an export license, the government would need a rather huge agency attempting detailed examination against some equally highly detailed and complex set of criteria. It would be highly demotivating to American companies with technology to sell or to transfer through licenses to foreign companies if the government approval and examination process were too long and costly. The moment an American company passes up an opportunity to exploit its technology offshore, foreign-based corporations will move in on the field. It would be even more disadvantageous to us if the rest of the world were to engage in liberal technology transfer from their countries into ours, using the transfer to broaden and build markets for their corporations and to exploit to the fullest any positions they might have on technology originating with them. The mutual stimulation and competition in the world market would cause their overall innovative capabilities to be enhanced. Meanwhile, our isolation would cause the quality of our efforts to slip.

Putting strong walls around the nation to stop export of technological know-how would not, and could not, be a unilateral act. It would be like any attempt at severe protectionist curbs to protect American industries; other nations would retaliate. We should remember that one out of eight American jobs is already based on manufacturing for export, half our grain crops go abroad, one-third of our farmland is committed to export crops, and a third of our companies' profits come from their activities abroad. Technology curbs are likely to resemble the trade wars of the 1930s that devastated our economy.

The lesser developed countries (LDCs) offer some special problems and opportunities with regard to technology transfer. On the one hand, those countries are most avid in their desire for advanced technology to come in from the developed countries. On the other hand, many of the LDCs argue that technological know-how, being merely wisdom, should belong to all, like the sun and the skies. Many of those nations are accustomed to receiving financial aid grants from the developed nations. If money itself can come for free, they reason, surely knowledge that could help them should be granted by the nation that possesses it, as a simple philanthropic act of civilized society.

However, to a private technological corporation, the technological know-how it possesses is one of its most valuable assets. Giving it away is pure charity and is no different from the corporation's contributing funds, products, or machinery without pay. The act would not be popular with shareholders who at the least would feel they ought to vote on each such gift and be able to take a personal income tax deduction for the charitable act.

In addition, pressed as they are with impossible social problems to solve, a constant requirement for LDC government leaderships is to find a scapegoat. The multinational corporation is often used for that role. An attack on a locally based American-owned company, accusing it of seeking to profit from its activity, is generally even more popular with the citizenry of the LDCs than with the American public. Typically, an LDC government desperately wants foreign corporations to come in and bring technology and jobs, so it seeks to make entry attractive for the corporation, often providing special tax advantages for a period of time and low-interest loans to provide some of the facility construction. At the same time, it creates an atmosphere that makes it difficult for the corporations to be successful in their relationships with the local population.

Whether we refer to technological transfer to the developed, industrial countries or to the LDCs, whether the fundamental issue in the transfer is national security or economic competitiveness in commercial world trade, technology export policy for the United States involves difficult-to-resolve dilemmas. The private sector plays a prominent part but the responsibility and decision-making role belongs to the government of the United States.

16

education for a technological society

Well chosen respective roles for free enterprise and government control are necessary if science and technology are to be used sensibly. But whatever schemes of organization and systems of decision making we might concoct, in the end the depth of the citizenry's understanding of the issues will set the degree of success attained in dealing with them. Some goals of our society may be quite obvious to the public, as may be some of the benefits and dis-benefits of pursuit of technological development. But on the whole, goals people want are expressible only through many overlapping layers of subgoals that often are not thought out and conflict with each other. Life and the people living it are so multifaceted we cannot easily tell what everyone is after, beyond such generalities as happiness, peace, and freedom. The big self-evident objectives on which the whole country might readily agree need to be broken down, interpreted, and interrelated, and this is an enormous intellectual task. Then, arranging to get what the nation desires, even when that appears both clear and possible, is so complex it requires both decentralization, to allow the maximum flexibility to individuals to do as they please, and centralization, to delegate power to a relatively small leadership group at the top. This group must be given the job of defining, deciding, and acting for us in many important areas. Without such arrangements, not enough will happen in our democracy to move us in the directions we want to go.

All this is another way of saying that in America how things work out will depend on the quality of perception and comprehension the public possesses. The private sector–government interrelationship is key in making things work. Making tradeoff decisions, soundly and on reasonable time schedules, is equally important. But we won't form that interrelationship correctly or balance tradeoffs right unless the people act with responsibility, determination, a sense of practical values, and enough know-how about the process that their participation will more often help than hinder. What the voters in the United States think and do depends on a broad spectrum of their individual backgrounds, experiences, prejudices, education, and interests. Let us try to advance our feel of how the citizens' influence toward intelligent application of science and technology can be enhanced by education. More particularly, we shall in this chapter examine the role of the university in providing a foundation for superior citizenship in a technological society. Of course, education alone does not make the citizen, and university education is not the only dimension of the educational process. However, the higher-educational system is a very strong pillar of U.S. society. Directly and indirectly, the universities of the country have a tremendous impact on how the voters act and have the potential of greatly improving their ability to think.

The colleges have established themselves as providing at least two important functions for the nation. Both are pertinent to our discussion. The first is to prepare the students for the world ahead. The second is broader and goes well beyond the classroom: to perform research, study important problems and opportunities of the society, extend the frontiers of understanding, and disseminate knowledge and ideas to the people at large.

As to the educational process, college graduates and those with partial university-level education comprise a strong block of American voters. They dominate in control of newspapers, television, radio, motion pictures, books, and magazines. They lead in business, industry, government, education, and the professions. What the college-educated voters of the United States wish to have occur, when that is possible, is very likely what will happen, especially since on most issues those with little formal education do not necessarily hold views in opposition to those with college training.

As to the second function, the nation already relies heavily on universities to study the more puzzling issues with which the society is concerned. What university professors believe and the results of their studies are widely published and seriously considered. The thinking time of many of the brightest, most knowledgeable people in all segments of the nation's activities outside of academia are consumed by urgent tasks of the moment in business, industry, government, and the professions. In contrast, the most skilled players in the university arena often operate in a cultural environment conducive to the philosophical, longer-range, and fundamental deliberations we badly need.

Universities all over the country are hard at work on the two functions just described. However, in the context of this book's subject, both activities greatly need expanding. As we continue the transition to an ever more technological society, we must improve the education of our young people and the supplementing of the education of our adult citizenry. If we don't, we shall be preparing them for the world of the past rather than the greatly changing world of the future, and we shall misuse or fail to use properly the tools of science and technology.

We teach English to all students from first grade on up, even to those with little aptitude for it. We teach them how to read, partly for survival, partly for information, partly to stimulate their thinking, and partly to broaden their cultural background. We teach them sometimes to speak and write English well and, in fewer instances, to spell correctly. We do all this not because we expect any but a very few to become professional writers or English teachers but because we want them to be able to communicate with others. No university would think it was turning out graduates with minimum qualifications if they could not handle the English language well enough to deal with communications coming to them by way of newspapers, television, and at least an occasional book.

Yet, in a nation that is a hybrid of a free enterprise and a government-controlled economy, we turn out many university graduates who have not even a minimal understanding of economics, the free enterprise system, and the relationship of private business to government. Even those who have had an introduction to economics in college rarely understand enough of practical economics for responsible citizenship. There is failure to grasp fundamentals, and misconceptions exist frequently as to important facts. A frequent assumption, for example, is that the profits of business operations are totally gravy, a reward over and above all costs and taxes. There is no realization that earnings rates must be above the threshold level of the cost of capital employed or no investment capital will be forthcoming. Few graduates leave college with an awareness of how low our ratio of investment to total GNP is presently, or that it is far lower than for other industrial nations.

A recent poll showed that the average new college graduate believes private industry's profits to be between 25 and 40 percent of sales, an error of 500 to 1000 percent. Too often, university graduates go along with generally less educated persons in assuming that new government expenditures can always be financed by a corporation tax increase. Soaking the corporations means the public will then be relieved of having to pay for such expenditures, they think, seemingly unaware that higher corporate taxes in the long term, and often in the short, have to be passed on to the public in the higher sales price of products. Many college graduates are as unable as a high school student to appreciate that inflation relates more than a little to money supply and the government's lack of control over it, and the state of the economy is connected to government spending and taxing.

But economics is not the only weak area in the typical college student's curriculum. Educated persons should know of their country, their world, and its people, the laws of nature and of political organization. Aside from the need for new specialists as the world changes, mature citizens, whatever their life's work, also must understand some of the fundamentals of science and technology. It is possible to obtain a degree from many institutions of higher learning with no feel whatsoever for the scientific approach and no knowledge of the existence of problems of selection in applying science and technology to our society's perceived needs. Most graduates do not recognize the concept that altering society involves alternatives, that for each change there are benefits and penalties, and that the challenge is one of balancing, matching, comparing, deciding, and acting. It will be ever more necessary in coming years to teach history, to use understanding of the past to prepare for the future. It would be advantageous also to develop new studies that are to the future what history is to the past. We need to accustom graduates to thinking of the possibilities ahead for the society.

The typical university introductory philosophy course dwells on what respected philosophers of old have had to say long before the world became as technologically oriented as it is today. Only the more modern and advanced courses attack the philosophical problems of the present world, and even these fail to cover realistically the impact of science and technology. Often only the past is discussed in class; the students mainly speculate about the future totally on their own.

That much of all scientific research and technological advance is now dominated by the federal government usually comes as a surprise even to engineering graduates. They go to work in industry after graduation thinking that applying science in industry depends on the free market. They are often astonished to discover the magnitude of government influence on the development of most technological products through patent policy, antitrust laws, labor–management relations, actions regarding inflation, regulations for safety and environmental control, taxes, and direct sponsorship of research and development in industry or at the universities.

Perhaps the greatest shortcoming of university programs, if their goal is to prepare the graduate to participate as a citizen in decisions on science–society relationships, is the lack of understanding of this relationship itself. Evidence of this is that the nation lacks a true profession dealing with the application of science and technology to the society. Considering the nature of so many of the nation's problems, and the way in which science and technology interact with all of them, it is not unreasonable to expect that, as in law and medicine, a recognized profession would exist and university degrees could be obtained in this specialty. Judging by its often stated definition, namely, "the application of science and technology to society's needs," engineering might appear to be that profession. But this is a myth. In both formal education and practice, engineering really is utilization of science, technology, and resources to design and build machines and systems. This is only a part of the full profession required.

The proper use of science and technology — their timely and wise application to help people with their problems, enhance their opportunities, and provide them with acceptable means to satisfy their requirements while maintaining a good balance of safety, health, and protection of the environment — is an endeavor of vast proportions. What we are speaking of is the matching of science and technology to social needs and progress, including everything from recognition of the need, articulating of options for filling it, and analysis of social, economic, and technological tradeoffs, right through to the planning, arranging, and actual implementing of the most sensible response. Such a task, requiring the combining of knowledge and ideas on so many technological, economic, social, and political fronts, is an intellectual challenge well above conventional engineering. The art of mixing together in a harmonious ensemble public value judgments, creativity, and technical analyses with workable pragmatic actions transcends the established expertise of any recognized profession. Yet the potentials match the difficulties. Also, the detriments to society of inadequacy of attack and execution are so great that professionalism in applying science and technology is a vital world need. Amateurism, with everyone in the act in a helter-skelter free-for-all, is in some respects inevitable as part of the operations of a democracy. But it does not have to be as bad as it is.

Most engineering educational institutions require their students to include courses in the humanities and social sciences. But this practice has been largely to apply a cultural veneer in an effort to make the engineering graduate a broader person. It has not come about out of a recognition that understanding the way our society operates is as important as understanding physics, if the student is to aspire to professionalism in applying science and technology to the society. If we trained our physicians the way we do our engineers, we would first conceive of medicine narrowly as involving (if the reader will forgive a slight exaggeration) the application of drugs and knives to the human body. Then we would proceed to teach medical students all about drugs and knives. We would also give them a few short courses on the human body—as a cultural bonus, not because of an imperative that says knowledge of the human body is vital to their professional work. We would then send the young medics out to apply the acquired tools (the knives and drugs) to the society, that is, to the human bodies that come their way.

If engineering were really the profession of applying science and technology to society, engineers would have to spend as much time in college learning about—and in their profession, working and dealing with—the society as about the tools they plan to apply to that society. While practicing engineers today do not on the average have a knowledge of society greater than that of many other professionals, such as lawyers, physicians, educators, businesspeople, politicians, or those selling insurance, neither are these other professionals practitioners in integrating science and technology with society's wants and needs. We have a missing profession.

This has been recognized by a very few universities. They are now offering hybrid courses in which science and technology, government, economics, and sociology are intermixed to create a new kind of interdisciplinary graduate. This is a beginning, but only a beginning, because graduates who are half physicist and half political scientist are not necessarily, and may not even think of themselves as being, members of a profession that applies science to the society. The requirement is for a much faster solidifying and developing of the idea of a professional activity that relates science and technology to national and world needs.

Aside from meeting the requirement for professionals, and educating to include appreciation of the challenges of the technological society, much more could be done to direct that society in the right directions if university research programs were adequately broadened. Of course, research in science and technology is of critical importance to the nation. We have already covered the need for the government to do a better job here, especially to arrange a more generous, longer-term, less bureaucratic approach. However, there is much more research that universities should do for the nation, beyond explorations in the physical and biological sciences.

The concept that in America we are a hybrid society, part government-controlled and part free enterprise, should be researched and clarified. Study is needed on the organizational approaches to the most important problems where science and technology are involved, such as energy, inflation, economic competitiveness, security, and improved transportation, communication, health, and environmental protection. For these areas, roles and mission assignments for the government and the private sector deserve more deliberation. The university environment is right for part of these studies, for contemplative and innovative discussion, for calm and objective consideration of all of the facets and the way they interact, with no pressure resulting from a scheduled due date for delivery of the results.

Basic to understanding better the relation between free enterprise and governmental control is another question also deserving research: the tie between free enterprise and overall individual freedom in America. We can justify time spent inventing how to use both the private sector and government in appropriate confrontations or teamings if we believe that free enterprise should be here to stay in America—not here to grow and be all-encompassing, but also not to disappear leaving the nation under total government control. If we eliminate the free enterprise role and look to the government to dominate the development of science and technology tools, we might gradually find the government in control of everything else. This, presumably, we would not allow, once we were to recognize it.

Consider the following chain. Free enterprise cannot be sustained without sufficient financial returns on private risk investment; if the government, with acquiescence by the voters, creates such conditions that the private sector cannot realize adequate returns, the free enterprise system will fail; by default, government control of the applications of technology will then become the nation's organizational pattern; this will result in government control of production; this will require the allocation by government of all materiel and human resources; this will lead to a totally government-controlled society with assignments of jobs and areas in which to live; ultimately, this equates to loss of all individual freedoms. Is this chain realistic, serious, possible? What governs the probability of its becoming an actuality? What is the tie, if any, of that important societal value, freedom, to technological advance and to the government's influence over it? These are but examples of basic socio–technological issues deserving attention by academic researchers.

We have been discussing education and the university's role based on very demanding, perhaps idealistic, ambitions for what higher education can accomplish. We have talked of the superior forming of the minds of the young and about elevating the citizens' ability to participate in decision making. However, we should ask now some narrower, more focused questions: How can the nation's skill in technological innovation be enhanced? Can the university be a stronger component in laying the groundwork for a higher quality of technological creativity? We should look specifically at those areas where growing technological inferiority is hurting America in the continuing economic and security competition with other nations. If we see ourselves falling behind in some field of technology that is important, we should ask how we can improve our competence in that specific field.

While we shall now discuss this with emphasis on the university contribution, it should be understood that both industry and government must be involved in any modifications and strengthenings of universities in matters relating to science and advanced technology. For one thing, the university cannot be expected to find the finances for substantial changeovers or expansions without aid from government or industry. It is very unlikely that huge increases in private individuals' and foundations' philanthropic contributions to universities will occur in the future. What universities need to do, particularly if they involve themselves in major changes, will require more than sympathy from government and industry. The budgeting of both of these sectors of the country will need to be generously reproportioned for financing university affairs so as to make needed modifications possible.

Let us probe some of these modifications. One is in the field of engineering. We have already amply covered pure research in universities except in one respect. We have used such adjectives as *basic* and *pure* for university research, to cover those intellectual pursuits that broaden human understanding of the laws of the universe with no regard for the nature and timing of possible applications. There is also a need for applied scientific research, and for research in the basic approaches to engineering. Unlike typical industry R&D that is focused on specific product endeavors, these investigations need to be broad and involved with intellectual disciplines. The potential for results of proprietary value is small for this kind of research, so it is carried on in private industry at a very low level, small companies hardly engaging at all in such efforts. Well-staffed university engineering departments could be just right for this type of exploration and study.

As an example, many thousands of companies are interested in cutting, casting, and forging of metals. Innovation in processing metals depends in part on knowing the basic metallurgy, the internal structure and properties of the material. Industrial concerns do a certain amount of fundamental work in this area, but they select for the highest-priority attention those lines of effort believed most likely to lead to proprietary materials and methods particularly useful in making their products. The broader fundamentals, helpful equally to all users in the future, are only lightly researched. Here expanded university programs would help all.

Closely allied to materials and their processing is manufacturing technology. A product is made of pieces, and these ultimately of various raw and finished materials. The manufacturing process that puts it all together is the first firing line of real-life hardware experience where the specified performance characteristics of the product will be built in, if all goes according to plan. Data that tell how the product will be produced can originate only with engineers who understand both the product and the process of making it using people and machines. It takes a highly developed skill to synthesize a manufacturing process, relate it to the product's design in great detail, and then find clues on the basis of which the product can be improved and the production process made easier, cheaper, safer, and more reliable. The success of American products in the domestic and international market hinges greatly on carrying on this kind of technological manufacturing effort in a manner superior to that of competitors.

We examined earlier the growing concern about slipping productivity in the United States. To correct this problem, our engineering schools should have an interest in the subject greater than shown by present meager courses in the curriculum and equally rare pertinent activities of the engineering faculties. University research programs should track down the relationships of productivity to such factors as capital investment, worker motivation, research and development expenditures, safety and environmental standards, utilization of computer-based information systems, the inflationary environment, tax policies, equal opportunity laws, and all the other interacting factors we mentioned earlier in the book. Of course, for research to improve our understanding of how all of these matters interrelate, more disciplines than those of engineering and science will have to be brought in. Universities should have centers for study of productivity, with participation from scientists and engineers, economists, psychologists, sociologists, and business professors. Such an interdisciplinary group with a common interest in productivity should also include individuals with substantial background in government and international matters, because part of the purpose of the research is to understand the distinctions in productivity performance between the United States and other nations and the way government policies and actions set the climate for productivity gains.

Good university centers for productivity research would benefit all industry and aid the government. They could develop an improved foundation for knowing what we are measuring and talking about in productivity questions, considering alternative approaches and seeing which of them can yield maximum benefits at minimum costs. We need both industry and government participation if university programs in productivity are to be adequately supported and realistically effective in what they produce. A model for this already exists at Massachusetts Institute of Technology, where a center for productivity and manufacturing technology is now active along the very lines described. Other universities have beginnings. A dozen substantial centers and fifty smaller efforts could be justified for a meaningful enhancement of American universities' potential contribution. The universities should be the recipients of industry philanthropy specifically for these purposes, and industry should provide experts to associate with the university efforts. Experienced production engineers should be loaned by large technological corporations on a year's leave to participate full-time in the activities of the universities' centers.

The government should include generous financing for this kind of activity in its budgeting. Present efforts along this line are much too modest. After a long period of study on how to increase technical knowledge to reverse our slippage in production technology, the government has proposed to add only $20 million to the budget of the National Science Foundation to fund new university R&D efforts in the field. It would pay off heavily for the United States if this figure were more than doubled right away.

What we have said about productivity and about production engineering applies to certain other areas of engineering that are very weak in the United States and where the universities are part of that weakness. We referred earlier to the serious trend of our loss of export trade in mechanical equipment and systems, with Germany and Japan in particular taking this business from us. As an example, automatic devices (robots) are being developed to lower costs in manufacture and provide added safety by replacing human operators in hazardous operations. Japan today has some 13,000 robots in its factories. The much larger United States has only 2500. Japan has set up a $50 million R&D program with the objective of experimenting with completely robot-operated factories.

Typical American university engineering programs are dominated by the more glamorous science that recently has found its way into engineering: semiconductor phenomena, lasers, outer space research, controlled thermonuclear fusion, cryogenics, and biological engineering. Without in any way diminishing the courses of instruction and research projects of universities in these frontier fields, it should be possible for joint industry, government, and university leadership to invent ways to motivate students and create professorial expertise in the fields basic to our regaining strength in mechanical engineering. Our growing weakness in mechanical systems and machinery shows itself strongly in our failure to lead in the design of machines that handle manufacturing activities. When we allow ourselves to be weak in machine design we are settling for being inferior in manufacturing.

Along with manufacturing technology, good old-fashioned product design technology is often neglected in American engineering education. Faculty interest in the technique of design is rarely noticeable and student appreciation of the process is correspondingly low. We are speaking here of mundane, grubby engineering — the designing of a product so that it will meet given specifications, cost less to produce, use a minimum of critical materials in short supply, be reliable in performance, and require reasonable maintenance in the hands of a customer.

If our goal is superior technological innovation, this can come about in part if innovation and creativity are probed and encouraged in American universities in all areas, not just in engineering. Most higher education rarely plants the seeds to grow creative ability, and when it does, little nourishment is applied to make it blossom. In fact, college course work often suppresses originality. In engineering education the students spend most of their time working problem examples by the specific methods which have been explained in class and in the textbooks. Through lectures and laboratory experiments, they learn how existing machines and systems work. They are taught prescribed ways to analyze certain situations and how to use a group of already developed analytical tools. They practice manipulating the quantitative formulas relating the various phenomena with which they will have to deal in existing engineering systems and devices.

All this is properly categorized as training in analysis, not synthesis or invention. The students not only fail to get an introduction to design — the finding of a practical, real-life solution to a bundle of requirements from performance to cost, size, weight, appearance, and customer satisfaction — they also get no introduction to the concept of being innovative. For that they would have to be offered frequent opportunities to take situations for which science and technology might offer solutions and then try to invent those solutions.

We should not confuse research work carried on to earn a Ph.D. with releasing and nurturing the innovative talent of the Ph.D. candidate. Some research work may stimulate creativity in the graduate student, but mainly what is sought by a Ph.D. dissertation is an investigation into the previously unknown, unmeasured, unanalyzed, uncorrelated, or uncategorized. The resulting, freshly uncovered information, to qualify as a Ph.D. dissertation, must be new in the sense that it has not been set forth yet by others, and it must be of significance in the field. But it need not stem from innovative thinking and that it do so is usually not a dominant requirement. Many Ph.D.'s, even very successful ones in science and engineering, would readily admit to being not very inventive or creative. On the other hand, a successful inventor has to be, first of all, inventive.

But we know there is more to technological superiority for the United States than technological innovation. Also, it is even true that there is more to technological innovation than creativity in the technology itself. Thus, aside from technology, university goals need to include the enhancing of student originality in approaching problems and challenges more generally. It is probably easy to stimulate a random flow of new proposals from college students whether the field be technology, sex, government, business, drugs, or lifestyle. But we are speaking of innovation that will pay off, of inventing to fill real needs, of matching ideas and wisdom to the requirements of the society. The problem is to encourage the forming of good ideas — ideas that will really work. This requires that the effort go beyond spurring original thinking, although that is a necessary requirement and a beginning. Useful innovation must derive from a superior understanding of both the problem and the criteria for a sound, acceptable solution.

At present, the creative effort of university students is more likely to be shown and brought out in extracurricular affairs. The existing curriculum does not go directly to the intellectual processes of innovating or to analysis that can guide and judge creating when attempted. Research programs and courses of study specifically intended to develop worthwhile innovative talent, whether of technological or non-technological character, are rather uncultivated fields as yet. If at the university level we are not yet able to deal with the underlying intellectual processes, research programs are necessary before courses of instruction can be constructed with confidence. However, experimental courses to stimulate and encourage creativity could be tried and these might even be part of the research process.

Other nations do not yet appear very active either in university programs dealing with innovation. Rarely, anywhere, is the creative process regarded as ready for laying out as an intellectual discipline. If, through avid interest and an innovative approach to innovation, we were to create new university programs in the subject, the results might be disappointing. Since such programs have hardly been tried, we really don't know. They might turn out to be remarkably productive.

17

what could happen; what should instead

What does it all add up to? This promise of exciting benefits from our extension of technology accompanied by our hesitancy in bringing about such gains because of new appreciation of potential negatives. This inability to organize properly to weigh the technological society's options, compare the good with the bad, and then make timely, balanced decisions. This confusion as to the roles of government and the private sector when we know perfectly well we need the involvement of both. What will happen to us in the period ahead? It is possible to be lucid in answering this question if we choose to be pessimistic, which is exactly what we must be if we assume present trends will continue. But we should not allow that. To see why, let us postulate precisely that sad situation.

If nothing different is done, we shall continue to see a high rate of inflation in the future for the same reasons we have it now. The nation's ratio of investment to consumption will stay low. It may even move downward. We shall go on countenancing high government expenditures and high taxes. Our research and development budget, like our capital investment, will be a decreasingly smaller portion of the GNP. The innovators in the private sector will do less innovating year by year, there being less financial support for their creativity. The low positive cash flows in technological industry, forced by the continuing inflation and the poor real returns on investment, will discourage risk taking. With neither capital investment nor R&D rising, productivity will continue to drop. Every year will see fewer new technological companies formed and the large established corporations will find it increasingly difficult to raise equity capital, the low stock market deservedly reflecting their mediocre earnings and dividends. They will increasingly be over-borrowed, a situation they will recognize as a growing danger, so they will become ever more conservative.

The real growth rate of the United States economy will be essentially nil. The average American will have less discretionary funds and living standards will drop. Energy supplies will diminish as petroleum-based fuel becomes less available and more expensive while stalemates will be the rule on development of significant energy alternatives. With all of this, there will be aggravated social instability, the disadvantaged suffering most and the middle class increasingly vocal about their dissatisfactions. The government will be under more severe pressure to act and will do so in a political, not rational, way, going a bit in every direction. With ever more government initiatives, a larger fraction of them will be the wrong actions and the private sector and the government will widen their adversary relationships. The public will become even more bewildered than now as to the proper roles and missions for government and free enterprise.

In this growing atmosphere of precariousness and confusion we shall make little headway in trying to arrange for objective tradeoff analyses to balance out benefits versus risks in our technological selection process. In fact, we can assume ever more persistent government control and over-regulation, including price controls, allocations, rationing, and burgeoning rules covering more aspects of life. These conditions will cause Americans to feel less secure. The rest of the world will observe this insecurity and share it. The United States will make an increasingly less significant positive contribution to world economic and social stability.

But this extreme forecast, obtained by merely extrapolating the bad trends of the 1970s, is not a given for the 1980s and beyond. In fact, even if we assume no overtly strong corrective action to arrange a superior state of affairs, there is a basis for challenging this pessimism. Some good things will happen without our consciously taking steps to promote their occurrence.

First of all, uncontrolled pessimism underestimates the impact of scientific discovery and technological advance. Even if we continue to be far from perfect in America in our selection process and the providing of motivation for investment in innovation, there will still be countless scientists and engineers scheming away in their laboratories and at their drafting boards. Valuable innovations will be forthcoming—not so large a number as we could ideally invent, but many, nevertheless. Through incorporating new ideas we surely shall produce food for ourselves and the world of higher quality and quantity each year. Some areas of technology that cry out for development will incite strong motivations and promise high rewards compared with investment risks, no matter how wrong government policies may be. For instance, a mint of money is there to be made by an auto manufacturer that leads the pack with real and sound innovation.

The United States will remain the single largest integrated economy and that will give us an edge in establishing new products. We shall make continued (and probably some amazing) advances in information technology, as exemplified by the microprocessor, the microminiaturized computer, which will increase our productivity in information handling, cutting expenses in almost all tasks we do. This will compensate for rising cost owing to inadequate investment in other dimensions of our operations where productivity may sink further. If we are not going to be as brainy as we should be in organization and decision making for the kind of technological society in which we live, we might be especially aided by synthetic brainpower. Matching human beings and electronic machines to create smart partnerships should represent a marked improvement in how we attack every problem and the efficiency with which we handle all routine informational chores.

Our military expenditures will probably increase as a percentage of the GNP over the next decade, and this could be extremely pertinent and beneficial to our leadership position in information technology, especially in computers and communications. This is because the right way for us to enhance our military strength in weapons systems is through superior communication, command, control, and overall utilization of our weapons systems—functions that depend on superior information technology.

The nation's energy programs may remain confused, to our disadvantage. We may go off half-cocked with expensive and largely useless programs. However, a great many competent scientists and engineers are going to be working on energy alternatives anyway, amid all the muddle of inconsistent sponsorship. It would be overly pessimistic to assume that no breakthroughs will come of this. Something very useful will surface, and in probably more than one field of energy. New techniques, involving everything from solar and nuclear approaches to improved discovery techniques for oil and gas, could upset pessimistic predictions on the energy front, reasonable as they now seem. We have hardly scratched the surface on energy conservation. If necessity is the mother of invention, this should apply to our inventing detailed changes in our way of life so that it can be interesting, exciting, satisfying, secure—and lower in energy use—by ingenious modifications in the how, what, and where about everything we do that uses energy. Looking back over the century, we see that again and again technological developments have altered the rules and required the discard of earlier assumptions. It would be foolish to imagine that we have come to the end of this course because the matching of technology to the society is now more complex. An innovative trend, even when slowed by all the limitations and bottlenecks we have listed, is still a trend. Even as the society has to change to use less energy per person, like it or not, we are bound to discover new ways to create additional energy supply.

One city or another might accomplish something approaching the spectacular in creating improved technological services. Advances may occur in urban transportation, energy distribution, waste disposal, water purification, and pollution control. One reason why urban services are unsatisfactory is that they are difficult to straighten out; there are too many interacting and opposing factors to consider, too many semiautonomous, selfish groups to deal with. But this means everyone will hear about and take notice of achieved success by any one region of the country if such a minor miracle should occur. A city that solves or appears to be solving a problem that almost everyone else is experiencing will find itself a celebrity town, its approaches emulated. When things get worse, Americans at some point begin to develop a resolve to do something about the evident problems. But they need leadership to point the way. If any city, through luck or unusual diligence or competence in its management, latches onto a more sensible approach to urban transportation, for instance, many other cities will be more ready to follow its lead than our pessimistic scenario might suggest.

No matter how much we handicap our use of the tools of science and technology, they will continue to be available, if dulled. Some scientific discoveries and technological breakthroughs will occur despite all our ineptness in organization and decision making in America. We are going to learn how to produce for our needs with some degree of lower pollution and hazards, lower prices and less call on resources. We shall discover some, if only a few, substitutes for materiel in short supply. The world will need America's food and this will bring dollars back that we will need to go on sending out of the country to pay for the products made better or more cheaply by other nations.

Progress in science and technology that turns out to be significant, even if we fail to provide high motivation for it, will not be the only force militating against a pessimistic scenario for the future. Free enterprise is another source of optimistic trends. Beneficial influences and some countering of bad trends will simply arise automatically out of market operations. The free market in America, even in the worst-case scenario with which we opened this chapter, is far from dead. Although weakened, free enterprise is so hard to kill in America that it will live a determined life of its own, partially defying and in other ways simply ignoring the generally non-integrated, often merely accidental, throttling down of it that occurs out of the workings of our society. To put it crudely, if there is a chance of making money, and doing so is not yet completely against the law, then someone will find a way to do it.

Some technological advances, a few perhaps based on fundamental, unpredictable scientific discoveries, will give rise to products so economically and socially desirable that, despite the scarcity of investment funds, they will be found somehow to back up the attractive implementations. Since one idea stimulates another and competition can be generally stimulating, it would be wrong for us to lay out a forecast of the future that overlooks the free enterprise dimension and assumes an enormous wet blanket of government control over all of the nation's sparks of activity.

Another powerful source of leverage toward progress is the international dimension. What happens to America depends increasingly on the policies of other nations, on the problems and pressures they experience and the contributions they make. Their actions can be beneficial for us in part and can make for a fuller utilization of science and technology by the world at large. For instance, if France and Germany, ignoring the American stalemate on nuclear energy, go that route all-out and successfully, it will diminish their competitive demands for petroleum. It will ease up the energy supply problem for the rest of the world, including America. If Japan leads in quality in manufacturing, high reliability in the product, and high productivity in the factory, it will set a world standard that we shall be driven to emulate. The alternative of accepting inferiority in design and production, when this is finally understood, will be unacceptable to Americans.

We remind ourselves that in the years ahead half or more of the good ideas in science and technology are bound to originate outside of the United States. Our ability to utilize science and technology to our advantage thus will depend to a great extent on our alertness and ability to bring in and adapt to our needs the advances that start outside of our boundaries. We can double every benefit—ways to extend or replace resources or lower production costs, new alternative energy sources, improved transportation and communication, better health care techniques — if we incorporate the advances in other countries. Even in our most pessimistic scenario, we should assume fairly open world trade. This means the advances attained elsewhere will be more or less thrown at us. Even if we bungle badly in shaping the American environment for innovation, we can still gain somewhat from the innovation of others.

Admittedly, in the worst possible world ahead, the contest we described earlier on the international trade front between a completely free world and an isolationist, barrier-ridden protectionist one would be won by the latter. Wars or imminent wars and a high degree of political, social, and economic instability everywhere in the world would work against the transfer of ideas and the ability of the world as a whole to gain from scientific and technological advances wherever they might occur. But the extreme of a halt to world trade is illogical to assume because if it should be about to occur the nations probably would have reached such a degree of disharmony that they would destroy society in a nuclear war first. Putting that possibility aside, for consistency we have to put emphasis on the economic and social value to all the world of free trade. It is sounder to assume that a substantial degree of such openness will manifest itself in the future.

We are concerned about the United States, but no other nation will be without its own set of severe problems. With time, it will be seen by most that the most effective approach to many of their problems is to negotiate and trade with others to achieve the maximum benefit from the ways in which countries can help one another. On the science and technology front, cooperation ensures that each nation will be in a stronger position to optimize the allotment of its limited natural and technological resource. Thus, the wave of the future might be exemplified well by Japanese and American corporations creating joint ventures to supply world markets in the most efficient way possible, the Japanese contributing what they are in best position to supply and the United States partners doing likewise. Proceeding in this way, multinational entities based in various nations around the world may develop strong, supportive positions.

It is in the nature of technological advance that most fields of endeavor require assembling interdisciplinary talents, the superior approach involving a combination of knowledge in various fields. A system solution to a problem or need, an answer that integrates harmoniously all of the various segments, will win out over segmented attacks. For instance, to handle electronics funds transfer in the United States, the architecture of the system must recognize how retailing operates here, how the industry and the American consumer want to relate to the new system. It must take account of the requirement for adherence to government regulations as to privacy, standardization, and fairness. Putting all this together requires skill in systems engineering assembly right here in America, but the hardware could well consist of pieces from Japan or Germany or France as well as those built in the United States. Similarly, for European or Asian or South American nuclear installations, parts of the system might be manufactured in the United States, incorporating new ideas discovered and developed here: other parts of the entire installation might come from other countries that have led in other segments. To achieve such optimized integration, bringing advanced technology from everywhere to bear on the local needs anywhere, obviously requires international technological organizations. This leads not only to stronger multinational companies, but to joint ventures of national and multinational companies.

In this final chapter, we have first looked at a worst scenario, one in which we postulated that all of the bad American trends would continue without change. We next indicated that a combination of three factors — science and technology, America's weakened but still very much alive free enterprise system, and international forces — might yield unplanned benefits and move us substantially away from that worst scenario. We now must add what could happen if we do the things we should do.

Eliminating every bottleneck to progress in the United States is asking too much. The problems are too severe for that and we do not have all the answers. Inflation is a good example of this. Inflation is probably the most penalizing factor limiting every aspect of national progress including sound and favorable advance of science and technology, but curbing it is a long and tough job. However, there are other aspects of our situation that we can remedy, if not easily and rapidly, then at least by reasonable diligence and in not too long a time.

The first to be mentioned is improvement of our examination of the negatives of technological advance, our procedures for comparing benefits and dis-benefits and, finally, our decision making intended to arrive at prudent balances. Here the course outlined in this book is recommended as a practical step: improving government regulation, enhancing the ability of the government to thoroughly investigate the potential negatives of our technological society in a timely fashion. If we implement the approach described in earlier chapters of setting up a strong unit in the government, SHEP, committed to safety, health, and environmental protection, with investigation alone, not decision making, as its mission, this would go a long way toward stopping wasteful over-regulation.

This would also require the parallel creation of the decision boards outlined in the book to perform tradeoff analyses and make the decisions on which technological implementations should be allowed to be launched or continued, banned or altered. Again, this does not have to be done for every avenue of life in which science and technology are factors. But it can be done for some right away. For example, we could establish immediately a decision board for energy. We could set up a decision board for occupational safety and hazards and another for environmental pollution. Again, these would not be perfect, but setting up these boards would be a great step forward in improving the regulatory process and would emphasize an absolutely indispensable government decision function, one only the government can handle.

The other and equally important step we can take is to improve the government–private sector relationship. The media, universities, government leaders, business managers, and people in the street should espouse the idea that ours is a hybrid society, part free enterprise and part government-controlled. All should disavow the two extreme views: one, that the government is a useless, inefficient bureaucracy that must be kept out of everything; and the other, that free enterprise equates to big profit-seeking business that undermines the society, so the government alone must control everything. If we can take this step, we need quickly to add another one. We should not look for a single, simple solution that tells us what the government and the free enterprise roles are. Instead, we should understand that life is too complicated to be handled by one universal formula for division of responsibilities and missions between the two sectors, government and private. Instead, the roles and missions have to be determined by the issues. Once we buy this, we shall surely be on the road toward assigning the right roles for the government in urban transportation, energy, communications, and the rest, and we shall see the free enterprise sector freed to pursue its role in technology advance.

Of course, if at the same time we can find answers to inflation and national security, unemployment and fairness, then science and technology can support and accelerate society's rise to a truly golden age, one in which science and technology are used to the fullest for the benefit of society.

index

Agriculture, *see* Food and nutrition
Airplane navigation, 19 – 20
Airplane technology, 79
Alaskan natural gas pipeline, 81– 82,
    107
Alaskan North Slope oil, 101
America's future, 12 – 40, 281 – 288
    automobile, and, *see* Automobiles
    domestic oil production and, 32 – 33
    food and nutrition and, *see* Food and
      nutrition
    geothermal energy and, 33
    health care and, *see* Health care
    legal profession and, 39 – 40
    nuclear fusion and, 31 – 32
    oceans and, *see* Oceans
    solar energy and, *see* Solar energy
    space technology and, *see* Space
      technology
America's status in science and
    technology, 41 – 62
    advances by other countries and,
      51 – 52
    applied research and, 46 – 47, 49, 53
    capital investment and, 49 – 51
    currency exchange rates and,
      45 – 46
    depreciation and, 54
    engineers and scientists and, 46, 47
    government and, *see* Government;
      Government regulation
    inflation and, 42 – 43, 53*n*., 54
    machinery and auto sectors and, 44
    military technology, 52
    new, small, technological corpora-
      tions and, 57 – 58
    patents and, 51
    private sector and, 53 – 62
    productivity of industry and, 41 – 43,
      45, 48 – 49

    pure science and, 53, 55
    relationship between universities and
      industry, 46
    research and development expendi-
      tures and, 45, 46 – 49
    return on investment, 54 – 55
    risk taking and, 55, 56, 58 – 59
    space satellite competition and, 44,
      53
    steel industry and, 42, 56
    summary, 68 – 70
    taxes and, *see* Taxes
    technologically intensive industries
      and, 44 – 45
    trade balances and, 43 – 44
    "Yankee ingenuity" and, 56
Antibusiness sentiment, 73
Antitrust laws, 83 – 88
    AT&T and, 87 – 88, 224 – 226
    foreign operations and, 242 – 245
    side effect of, 88
Aquaculture, 30
AT&T, 87 – 88, 224 – 226
Automobiles, 22 – 26
    conservation and, 164, 169, 173 – 176
    diesel engines and, 24 – 25, 207
    fuel cells and, 24
    future stakes in, 25 – 26
    gas consumption and, 22 – 23
    government and, 26, 82, 205 – 209
    government purchase of high-MPG
      cars, 150
    imaginary look at early years of,
      127 – 128
    innovations in, 92
    international comparisons, 240 – 241
    mass transit compared to, 180
    microprocessors and, 23
    modern advances in, 23 – 24
    new materials and, 23

pollution and, 22 – 25, 205 – 209
research and development and, 24
safety of, 206, 208
tradeoffs involving, 206 – 207
"Automotive Innovation and Productivity Act of 1979," 92

Basic research, 75 – 79, 90, 222 – 223, 275
Beuche, Arthur, 132n.
Biomass, 29 – 30, 148 – 150, 191 – 192
Blood substitute, 35
Brown, Harold, 53n.

Calvin, Melvin, 30
Canadian tar sands, 101 – 102
Cancer causing agents, 96 – 97
Capital investment, 49 – 51
Carter, Jimmy, 123, 130, 136, 137, 154, 160, 241, 263
CAT scanner, 36
Cattails, 30
Chemical industry, 59
Civil Service System, 88 – 89
Clean Air Act, 103, 105
Coal, 122
    conservation and use of, 171 – 172
    see also Synthetic fuel
Cogeneration, 166
Computers, 37 – 40
    see also Electronic information technology
Concorde, 49
Congress, 103 – 104, 115 – 116
Conservation, 150 – 151, 162 – 176
    allocation of energy and, 169 – 171
    auto MPG and, 164, 169, 173 – 176
    coal and, 171 – 172
    cogeneration and, 166
    community, 164 – 165
    corporate, 164, 165
    costs of, 165 – 166
    effect of, 163 – 164
    electric motors and, 168
    examples of, 162 – 163
    free market and, 168 – 176
    gasoline price and, 169 – 170, 173 – 174
    government and, 166 – 176
    information exchange and, 163

light bulbs and, 168
    wood burning and, 171
Currency exchange rates, 45 – 46

Data processing, see Electronic information technology
Decision boards, 111 – 116
    combining agencies and, 113 – 114
    Congress and, 115 – 116
    courts of law and, 112 – 113
    described, 111 – 112
    environment and, 205, 209
    as extension of presidency, 114
    Kemeny on, 115
    nuclear energy and, 139 – 142
    operation of, 112
    political nature of, 115
    responsibilities of, 114 – 115
    synthetic fuel and, 160
    tailoring of, 114
DeGaulle, Charles, 257 – 258
Department of Agriculture, 102
Department of Energy, 100, 157, 159 – 160
Depreciation, 54, 101
Diesel engines, 24 – 25
Disbenefits of technology, 95 – 116
    Alaskan North Slope oil and, 101
    California LNG controversy and, 106 – 107
    Canadian tar sands and, 101 – 102
    cancer causing agents, 96 – 97
    chemicals and, 96
    clean air "offset" requirement and, 109
    Congress and, 103 – 104
    decision boards and, see Decision boards
    East Coast refinery controversy and, 105 – 106
    energy versus environment and, 99 – 102
    Georges Bank controversy and, 106
    government standards and, 97 – 98
    herbicides and, 102
    OSHA litigation and, 107 – 109
    pesticides and, 102
    rearrangement of society and, 97 – 98
    regulatory agencies and, see Regulatory agencies

regulatory system and, 105
therapeutic drug industry and, 109 –
    110
tradeoff decisions and, *see* Tradeoff
    decision
as unavoidable side effect, 95 – 96
Disinvestment in backward industries,
    57
DNA, recombinant, 33 – 34

Ecology, 6 – 7, 10
    see also Environment
Education, *see* Universities
Electric power, 120*n*. – 121*n*., 122, 124 –
    125
Electronic information technology,
    210 – 233
    AT&T and, 224 – 226
    basic research and, 222 – 223
    capabilities of, 212 – 213
    citizen access to, 228 – 229
    communications and, 213, 224 – 233
    control of production and distribution
        and, 215 – 219
    Electronic Funds Transfer (EFT) and,
        213 – 215, 231
    free enterprise and, 221 – 233
    future and, 211 – 212
    government and, 214 – 215, 216,
        219 – 220, 221 – 233
    implications of, 210 – 211
    international scene and, 227 – 228
    mail service and, 231 – 232
    monopolies and, 224 – 226, 229
    optimistic view of, 219, 220 – 221
    pessimistic view of, 219 – 220
    privacy versus public access and,
        230 – 231
    satellites and, 19, 226 – 228
    two-way TV and, 217, 220 – 221,
        229 – 230
    United States market and, 223 – 224
Energy: agriculture and, 189 – 192,
    195 – 197
    Alaskan natural gas pipeline and,
        81 – 82
    Alaskan North Slope oil and, 101
    Canadian tar sands and, 101 – 102
    conservation of, *see* Conservation
    cost of, 66

free market system and, 153 – 154
government involvement in, 81 – 82
government's erratic policy on,
    99 – 102
reduced use of, 67
tradeoff concerning, 99
see also specific forms of energy
Energy Mobilization Board, 141, 160
Engineering manpower, 46
Environment, 198 – 209
    automobiles and, 22 – 25, 205 – 209
    cancer causing agents and, 96 – 97
    decision board and, 205, 209
    effects on, of non-nuclear alternatives,
        122
    employment of science and technol-
        ogy and, 200 – 204
    Environmental Project Office and,
        202 – 205
    EPA and, 204 – 205, 209
    evils of technology and, 6, 7
    fictitious examples of dealing with
        problems of, 198 – 200
    government role in, 200 – 209
    leadership and, 200, 202
    present system of dealing with,
        203 – 204
    private enterprise and, 201 – 204
    productivity tradeoffs involving, 63,
        64, 65
    SHEP and, 203, 205, 209
    tradeoff between energy supply and,
        99 – 102
    see also Disbenefits of technology
Environmental Protection Agency (EPA),
    59, 100, 103, 204 – 205, 209
Europe and Japan, 8, 9, 34, 35
    antitrust attitude of, 85 – 86
    application of technological ad-
        vances, 49
    capital investment in, 50
    currency exchange rates and,
        45 – 46
    economic controls and, 93 – 94
    electronic information technology
        and, 223 – 224
    engineers and scientists of, 46
    national security issues and, 252,
        255 – 258
    nuclear energy in, 134 – 136

productivity in, 41 – 42
railroad technology in, 86
research and development expendi-
    tures of, 45, 47
scientific and technological de-
    velopments by, 51 – 52
space satellite programs of, 44
tax policies of, 57
trade balances of, 43
Extending human brainpower, *see* Elec-
    tronic information technology

Federal Bureau of Investigation (FBI), 111
Federal Communication Commission
    (FCC), 225, 229, 231 – 232
Fermentation microbes, 30
Food and Drug Act, 104
Food and Drug Administration (FDA), 59,
    103
Food and nutrition, 16 – 19, 185 – 197
    advances in, 17
    allocation of water, land, and energy
        and, 195 – 197
    biomass and, 29, 30, 191 – 192
    early predictions for, 16
    ecology and, 17 – 18, 102
    energy use and, 189 – 192
    government role in, 185 – 197
    large scale farms and, 197
    leadership problem and, 18 – 19
    microorganisms and, 191
    national policy and strategy and,
        186 – 188
    nitrogen fixation and, 18
    photosynthesis and, 18
    potential of U.S. position in, 16, 18
    regulation of, 193 – 195
    research and, 188 – 193
    science of nutrition and, 189
    self-sufficiency of U.S., 16
    tradeoffs involving, 67, 102
    weather prediction and, 192 – 193
Ford, Gerald, 74
Free enterprise and pervasive govern-
    ment, 71 – 94
    airplane technology and, 79
    antibusiness sentiment and, 73
    antigovernment sentiment and, 73
    antitrust laws and, 83 – 88
    automotive innovations and, 92

basic research and, 75 – 79, 90
business leaders and, 74
Civil Service-private sector rivalry
    and, 88 – 89
control over participation and,
    82 – 83
elimination and economic controls
    and, 93 – 94
energy and, 81 – 82
enormous size of investments needed
    and, 72 – 73, 85
environment for technological inno-
    vation and, 88 – 94
government financing and, 92 – 93
guidance for the interaction of,
    90 – 94
military research and development,
    79 – 81
patents and, 82, 83 – 85
presidency and, 89
public expectation of government ser-
    vices and, 72, 73
railroad technology and, 86
relationship of, 74 – 75
removal of government interference
    and, 71 – 74
scope of government involvement,
    82, 90 – 91
standards for health and safety and,
    93
summary, 94
tax immunities and, 91 – 92
telephony and, 87 – 88
universities and, 75 – 79, 90
Fuel cells, 24

Gasohol, 29
Geothermal energy, 33
Government: airplane navigation and,
    20
    antibusiness prejudice of, 88, 89
    automobile industry and, 26
    capital investment and, 50
    decision boards and, 111 – 116
    disinvestment in backward industries
        and, 57
    DNA regulation and, 34
    domestic oil production and, 33
    environment and, 200 – 209
    food and nutrition and, 18 – 19, 185 – 197

free enterprise and, *see* Free enter-
    prise and pervasive government
geothermal energy and, 33
health care and, 36, 39
mass transit and, 178 – 184
national security and, 250 – 267
nuclear energy and, 130 – 133
nuclear fusion and, 32
oceans and, 30 – 31
regulation by, *see* Government regu-
    lation
relations between private sector and,
    88 – 89, 239 – 249
research and development pro-
    grams of, 46, 47 – 48
solar energy and, 30 – 31
space technology and, 20, 22
subsidies, 153, 154
synthetic fuel and, 156 – 161
taxes, *see* Taxes
Government regulation: agriculture
    and, 193 – 195
bureaucracies and, 59, 60, 61
chemical industry and, 59
effects of, 59 – 61, 65
energy versus environment and,
    99 – 102
free enterprise and, *see* Free enter-
    prise and pervasive government
as handicap to productivity in-
    creases, 59 – 62
increased costs and, 99
inflation and, 72
negative results of, 65, 99
pharmaceutical industry and, 61
regulatory agencies and, 98 – 116
summary, 93 – 94
tradeoff decisions, *see* Regulatory
    agencies
tradeoffs involving, 63 – 64, 65
see also Disbenefits of technology
Gray, Harry, 28
Grayson, C. Jackson, Jr., 93
Greenhouse effect, 155

Health care, 33 – 39
blood substitute, 35
CAT scanner, 36
computers and, 37 – 39
DNA and, 33 – 34

drug administration, 37
genetic science, 33 – 35
government and, 36, 39
interferon, 34 – 35
measurement techniques, 35 – 36
microsurgery, 35
repair of sensory systems and, 36 – 37
ultrasound, 36
Herbicides, 102
Holdren, John, 152n.
Howell, David, 134
Human race, physical changes in, 1, 2, 3
Hydrocarbons, 29, 30, 35
Hydrogen, 29

IBM, 224 – 225
Indian Point power station, 124
Inflation, 9 – 10
government regulation and, 72
productivity and, 42 – 43
for research and development, 53n.
taxes and, 54
Interferon, 34 – 35
International dimension, 234 – 249
antitrust laws and, 242 – 245
auto industry comparisons in, 240 –
    241
benefits to America and, 237 – 238
corporation-government relations
    and, 239 – 249
desire for technology and, 234 –
    235
environmental controls and, 242
foreign competition and, 238
free trade versus protectionism and
    236 – 237
government policies and, 238 – 249
investment and, 247 – 248
joint ventures and, 242
maximized performance parameters
    and, 235 – 236
multinational corporations and, 239
nuclear power and, 241
overseas production and, 245 – 246
patents and, 243
taxes and, 245, 247
telecommunications and, 241 – 242
world market and, 235

Japan, *see* Europe and Japan

Joint Economic Committee of Congress, 42

Kemeny, John G., 115

Malnutrition, see Food and nutrition
Mass transit, see Urban transportation
Microprocessors, 23
Microsurgery, 35
Missing decision department, 95 – 116
Medicine, see Health care
Microorganisms, 191
Military research and development, 52, 79 – 81
  civilian applications of, 79 – 80
  free market and, 80 – 81
  nuclear weapons and, 81
  see also National security
Monopolies, 83 – 88, 224 – 226, 229
"Moon Treaty," 21
Multinational corporations, 239, 265

National goals, 4, 268
National Institutes of Health, 34, 189
National security, 250 – 267
  free enterprise and, 254, 263 – 267
  government and, 250 – 267
  hawks versus doves and, 255
  Japan and Germany and, 252, 257 – 258
  lesser developed countries and, 266 – 267
  NATO and, 255 – 258
  non-military technology and, 250, 251
  over-emphasis on, 251 – 252
  present organizational system of, 253 – 254
  risk tradeoffs and, 253 – 258
  technology transfer and, 258 – 267
National Security Agency, 230 – 231
NATO, 255 – 258
Natural gas, 147 – 148
Nitrogen fixation, 18
Nuclear energy, 117 – 142
  accidents involving, 123 – 124, 125 – 126
  alternatives to, 121 – 123
  breeder reactors and, 129 – 130, 132, 135, 136
  conflicting opinions about, 119, 120, 125, 128 – 129

decision board and, 139 – 142
defined, 120n. – 121n.
economic advantages of, 117
electric power and, 120n. – 121n., 122, 124 – 125
Energy Mobilization Board and, 141
environmental effects of alternatives to, 122
existing dependency on, 123
in foreign nations, 134 – 136
government role in, 81, 130 – 133
Indian Point power station and, 124
international position of the U.S. in, 241
Nuclear Regulatory Commission and, 141
nuclear wastes and, 131, 131n. – 132n.
nuclear weapons capability and, 118 – 119, 130
optimistic scenario of, 130 – 133
politics and, 120
possible abandonment of, 121 – 122, 133 – 136
present conditions and, 121
private sector and, 120 – 121
problems with, 118 – 119
proposal for decision making on, 137 – 142
results of closing existing plants, 123 – 124
risks and, 119 – 120
safety of, 125 – 130
safety requirements of, 117 – 118
St. Lawrence Waterway project and, 132
SHEP and, 138 – 142
small doses of radiation exposure and, 128 – 129
stalemate in question of, 121
Three Mile Island and, 120n., 123 – 124, 125 – 126, 129, 134
uranium and, 112n., 118
Nuclear fission, 31, 120n. – 121n.
Nuclear fusion, 31 – 32
Nuclear reactors, 31
  breeder, 118, 121n., 129 – 130, 132, 135, 136
  described, 120n. – 121n.
Nuclear Regulatory Commission (NRC), 124, 126, 141
Nuclear weapons, 81

Occupational Safety and Health Administration (OSHA), 60, 103
litigation concerning, 107 – 109
Oceans, 26 – 28
government involvement in, 30 – 31
inhibitions to investments in, 27 – 28
as source of energy, 28
as source of resources, 26 – 27
United Nations and, 27
Oil: American dependence on foreign, 121 – 122
de-controlled, 147
domestic production of, 32 – 33
pricing of, 101 – 102, 147
secondary and tertiary, 146 – 147
shale, 101, 102, 148
synthetic, see Synthetic fuel
Organization, 18 – 20

Patents, 51
antitrust laws and, 83 – 84, 243
government control of, 82
government ownership of, 84 – 85
small businesses and, 84
value of, 83
Pesticides, 102
Pharmaceutical industry, 61, 109 – 110
Photosynthesis, 18
Postal Service, 231 – 232
Press, Frank, 53n.
Productivity, 63 – 70
employees and, 66
energy and, 66, 67
government regulation and, 59, 65
improvement in goods and services and, 64
increases in, 9, 41 – 42
increase in quality of life and decrease in, 63 – 64
inefficiency and, 64 – 65
inflation and, 42 – 43
magnitude of, 42
measurement of, 63 – 65
research and development and, 45, 48 – 49
return on investment and, 66
tradeoffs involving, 63 – 64, 65, 67 – 68
training of workers and, 67
ways to improve, 45, 66

Quality of life, 63 – 64

Railroad technology, 86
Regulatory agencies, 98 – 116
combining of, 113
Congress and, 103 – 104
control over, 104
conversion of, to investigative agencies, 110 – 116
cost of regulation, 104
decision boards and, see Decision boards
litigation and, 107 – 109, 112 – 113
as negating agencies, 103, 108
tradeoff decisions and, 98 – 116
Research and development (R&D): agricultural, 188 – 193
applied, 46 – 47, 49, 53
basic, 75 – 79, 222 – 223
decreasing expenditures on, 9, 45
engineering of machinery, 47 – 48
engineers and scientists engaged in, 46
as fraction of GNP, 45, 47
government regulation and, 61
government sponsored, 46, 47 – 48, 75, 90 – 91
inflation rate for, 53n.
managers of, 56
military, 79 – 81
productivity and, 45, 48 – 49
Return on investment, 54 – 55
government regulation and, 61
productivity and, 66
taxes on, 56 – 57
Revenue Act of 1969, 58
Richardson, Elliott L., 27
Rickover, Admiral, 126
Risk taking, 55
comeback in, 58 – 59
government regulation and, 61
tax policies and, 56, 58

Safety, Health and Environmental Protection (SHEP) (proposed agency), 113 – 116
agriculture and, 194n., 195
environment and, 203, 205, 209
mass transit and, 183
nuclear energy and, 138 – 142
synthetic fuel and, 160

Safety of workers, 63, 64, 65
Satellite, *see* Space technology
Science and technology, 1 – 14
  accomplishments of, 5 – 6
  American leadership in, 8 – 9
  America's future and, *see* America's
    future
  America's status in, *see* America's
    status in science and technology
  antitechnology sentiment and, 6 – 7
  benefits of, 9 – 10
  confidence in, 4 – 5
  disadvantages of, 6 – 7, 10
  disbenefits of, *see* Disbenefits of
    technology
  expenditures of research and de-
    velopment and, 9
  future of, 5, 12 – 14, 152
  government involvement in, *see* Free
    enterprise and pervasive gov-
    ernment
  hypothetical example of disparity in,
    11 – 12
  increased use of, 7 – 8
  inflation and, 9 – 10
  international aspects of, *see* Interna-
    tional dimension
  knowledge explosion in, 2 – 3, 5 – 6
  limits to, 7
  national goals and, 4
  as only area of human advancement,
    1 – 2
  politics and, 7
  social-political progress and growth
    of, 1, 3 – 4
  technology slip and, 7 – 8
Social adjustment to technological
  change, 1, 3 – 4
Solar energy, 28 – 31, 151 – 153
  applications of, 28 – 29
  aquaculture, 30
  biomass as, 29
  cattails and, 30
  as controversial issue, 28
  exaggerated expectations for, 151 –
    152
  government involvement in, 30 – 31
  hydrocarbons and, 29, 30
  hydrogen and, 29
  ocean-thermal, 28

  subsidization of, 152 – 153
  wood conversion and, 30
Soviet Union, 52, 68, 135 – 136, 187, 219
  see also National security
Space technology, 19 – 22
  airplane navigation and, 19 – 20
  commercial fishing and, 21
  communication and, 19, 226 – 228
  economic return on investment of,
    21 – 22
  European competition in, 44
  food and nutrition and, 17
  future advances in, 20
  government and, 20, 22, 82
  international relations problems and,
    21
  "Moon Treaty" and, 21
  observation of earth and, 17, 19, 20 – 21
  organization and, 20, 22
  U.S. lead in, 53
Standard of living, 64
Standards, setting, 93
Steel industry, 42, 56
Synthetic fuel, 143 – 161
  biomass and, 148 – 150
  burning garbage and, 149
  Canadian tar sands and, 101 – 102
  coal gasification and, 144
  conservation and, 150 – 151
  controversy over, 143 – 144
  decision board and, 160
  defined, 143*n.*
  Department of Energy and, 159 – 160
  energy alternatives and, 147 – 156
  environmental controls and, 158 – 159
  free enterprise system and, 145 – 147,
    153 – 154, 156 – 161
  government direction and develop-
    ment of, 156 – 161
  government purchase of high-MPG
    cars and, 150
  greenhouse effect and, 155
  indecision about, 145
  natural gas and, 147 – 148
  negatives of, 154 – 156
  oil and, 147
  potential of, 143
  pricing of, 101 – 102, 147
  proposal for development of, 159 – 161
  shale rock and, 148

SHEP and, 160
solar energy and, 151 – 153
in South Africa, 144
Taxes, 50
capital gains, 57, 58
depreciation and, 54, 161
as discouraging investment, 56
foreign operations and, 245, 247
private sector research and, 54
research and development breakthroughs and, 91 – 92
on return on investment, 56 – 57
Revenue Act of 1969, 58
risk taking and, 56, 58
Telephony, 87 – 88
Therapeutic drug industry, 109 – 110
Three Mile Island, 120n., 123 – 124,
125 – 126, 129, 134
Trade balance, 43 – 44
Tradeoff decisions, 63 – 64, 65, 67,
98 – 116
avoidance of, 103
decision boards and, see Decision
boards
difficulty of, 105 – 107
litigation as a result of, 107 – 109, 112 –
113
requirements for unbalanced
decisions and, 104
solving problem of, 110 – 116

Ultrasound, 36
United Nations, 21, 27
Universities, 268 – 280
applied research at, 275 – 279

basic research at, 75 – 79, 90, 275
broadened research at, 273 – 279
dissemination of knowledge and,
269 – 270
economics and, 270 – 271
education process and, 269
engineering and, 272 – 273, 275 –
279
failures of, 270 – 275
innovation and, 279 – 280
philosophy and, 271
private sector-government relationship and, 268 – 269, 274 – 275
public comprehension and, 268 –
269
science-society relationships and,
272 – 273
Urban transportation, 177 – 184
auto use compared to, 180
benefits of, 177
calculation of feasibility of, 181
government and private involvement
in, 178 – 180
hypothetical example of, 177 –
178
proposal for, 181 – 184
Washington, D.C. subway and, 179

Weather prediction, 192 – 193
Workers, 63 – 67
safety of, 63, 64, 65
productivity of, 66
training of, 67

"Yankee ingenuity," 56